Bone Marrow and Stem Cell Transplantation

METHODS IN MOLECULAR MEDICINE™

John M. Walker, Series Editor

139. **Vascular Biology Protocols,** edited by *Nair Sreejayan and Jun Ren, 2007*
138. **Allergy Methods and Protocols,** edited by *Meinir G. Jones and Penny Lympany, 2007*
137. **Microtubule Protocols,** edited by Jun Zhou, *2007*
136. **Arthritis Research:** *Methods and Protocols, Vol. 2,* edited by *Andrew P. Cope, 2007*
135. **Arthritis Research:** *Methods and Protocols, Vol. 1,* edited by *Andrew P. Cope, 2007*
134. **Bone Marrow and Stem Cell Transplantation,** edited by *Meral Beksac, 2007*
133. **Cancer Radiotherapy,** edited by *Robert A. Huddart and Vedang Murthy, 2007*
132. **Single Cell Diagnostics:** *Methods and Protocols,* edited by *Alan Thornhill, 2007*
131. **Adenovirus Methods and Protocols, Second Edition, Vol. 2:** *Ad Proteins, RNA, Lifecycle, Host Interactions, and Phylogenetics,* edited by *William S. M. Wold and Ann E. Tollefson, 2007*
130. **Adenovirus Methods and Protocols, Second Edition, Vol. 1:** *Adenoviruses, Ad Vectors, Quantitation, and Animal Models,* edited by *William S. M. Wold and Ann E. Tollefson, 2007*
129. **Cardiovascular Disease:** *Methods and Protocols, Volume 2: Molecular Medicine,* edited by *Qing K.Wang, 2006*
128. **Cardiovascular Disease:** *Methods and Protocols, Volume 1: Genetics,* edited by *Qing K. Wang, 2006*
127. **DNA Vaccines:** *Methods and Protocols, Second Edition,* edited by *Mark W. Saltzman, Hong Shen, and Janet L. Brandsma, 2006*
126. **Congenital Heart Disease:** *Molecular Diagnostics,* edited by *Mary Kearns-Jonker, 2006*
125. **Myeloid Leukemia:** *Methods and Protocols,* edited by *Harry Iland, Mark Hertzberg, and Paula Marlton, 2006*
124. **Magnetic Resonance Imaging:** *Methods and Biologic Applications,* edited by *Pottumarthi V. Prasad, 2006*
123. **Marijuana and Cannabinoid Research:** *Methods and Protocols,* edited by *Emmanuel S. Onaivi, 2006*
122. **Placenta Research Methods and Protocols:** *Volume 2,* edited by *Michael J. Soares and Joan S. Hunt, 2006*
121. **Placenta Research Methods and Protocols:** *Volume 1,* edited by *Michael J. Soares and Joan S. Hunt, 2006*
120. **Breast Cancer Research Protocols,** edited by *Susan A. Brooks and Adrian Harris, 2006*
119. **Human Papillomaviruses:** *Methods and Protocols,* edited by *Clare Davy and John Doorbar, 2005*
118. **Antifungal Agents:** *Methods and Protocols,* edited by *Erika J. Ernst and P. David Rogers, 2005*
117. **Fibrosis Research:** *Methods and Protocols,* edited by *John Varga, David A. Brenner, and Sem H. Phan, 2005*
116. **Interferon Methods and Protocols,** edited by *Daniel J. J. Carr, 2005*
115. **Lymphoma:** *Methods and Protocols,* edited by *Timothy Illidge and Peter W. M. Johnson, 2005*
114. **Microarrays in Clinical Diagnostics,** edited by *Thomas O. Joos and Paolo Fortina, 2005*
113. **Multiple Myeloma:** *Methods and Protocols,* edited by *Ross D. Brown and P. Joy Ho, 2005*
112. **Molecular Cardiology:** *Methods and Protocols,* edited by *Zhongjie Sun, 2005*
111. **Chemosensitivity:** *Volume 2, In Vivo Models, Imaging, and Molecular Regulators,* edited by *Rosalyn D. Blumethal, 2005*
110. **Chemosensitivity:** *Volume 1, In Vitro Assays,* edited by *Rosalyn D. Blumethal, 2005*
109. **Adoptive Immunotherapy:** *Methods and Protocols,* edited by *Burkhard Ludewig and Matthias W. Hoffman, 2005*
108. **Hypertension:** *Methods and Protocols,* edited by *Jérôme P. Fennell and Andrew H. Baker, 2005*
107. **Human Cell Culture Protocols,** *Second Edition,* edited by *Joanna Picot, 2005*
106. **Antisense Therapeutics,** *Second Edition,* edited by *M. Ian Phillips, 2005*
105. **Developmental Hematopoiesis:** *Methods and Protocols,* edited by *Margaret H. Baron, 2005*
104. **Stroke Genomics:** *Methods and Reviews,* edited by *Simon J. Read and David Virley, 2004*
103. **Pancreatic Cancer:** *Methods and Protocols,* edited by *Gloria H. Su, 2004*
102. **Autoimmunity:** *Methods and Protocols,* edited by *Andras Perl, 2004*
101. **Cartilage and Osteoarthritis:** *Volume 2, Structure and In Vivo Analysis,* edited by *Frédéric De Ceuninck, Massimo Sabatini, and Philippe Pastoureau, 2004*
100. **Cartilage and Osteoarthritis:** *Volume 1, Cellular and Molecular Tools,* edited by *Massimo Sabatini, Philippe Pastoureau, and Frédéric De Ceuninck, 2004*
99. **Pain Research:** *Methods and Protocols,* edited by *David Z. Luo, 2004*
98. **Tumor Necrosis Factor:** *Methods and Protocols,* edited by *Angelo Corti and Pietro Ghezzi, 2004*

METHODS IN MOLECULAR MEDICINE™

Bone Marrow and Stem Cell Transplantation

Edited by

Meral Beksac

Department of Hematology, School of Medicine
Ankara University, Ankara, Turkey

HUMANA PRESS ✱ TOTOWA, NEW JERSEY

999 Riverview Drive, Suite 208
Totowa, New Jersey 07512

www.humanapress.com

This publication is printed on acid-free paper. ∞
ANSI Z39.48-1984 (American Standards Institute)

Permanence of Paper for Printed Library Materials.

Cover design: Karen Schulz

Cover illustration: Figure 2A superimposed on Fig. 2B from Chapter 11, "Identification of Bone Marrow Derived Nonhematopoietic Cells by Double Labeling with Immunohistochemistry and *In Situ* Hybridization," by Isinsu Kuzu and Meral Beksac.

For additional copies, pricing for bulk purchases, and/or information about other Humana titles, contact Humana at the above address or at any of the following numbers: Tel.: 973-256-1699; Fax: 973-256-8341; E-mail: orders@humanapr.com; or visit our Website: www.humanapress.com

Printed in the United States of America. 10 9 8 7 6 5 4 3 2 1

eISBN-10: 1-59745-223-8
eISBN-13: 978-1-59745-223-6

ISSN: 1064-3745

Library of Congress Cataloging-in-Publication Data

Bone marrow and stem cell transplantation / edited by Meral Beksac.
p. ; cm. -- (Methods in molecular medicine, ISSN 1064-3745 ; 134)
Includes bibliographical references and index.
ISBN-10: 1-58829-595-8 (alk. paper)
ISBN-13: 978-1-58829-595-8 (alk. paper)
1. Bone marrow--Transplantation. I. Beksac, Meral. II. Series.
[DNLM: 1. Bone Marrow Transplantation--immunology. 2. Bone Marrow Transplantation--methods. 3. HLA Antigens--immunology. 4. Hematopoietic Stem Cell Transplantation--methods. 5. Hematopoietic Stem Cells--immunology. 6. Histocompatibility Testing. W1 ME9616JM v.134 2007 / WH 380 B7088 2007]
RD123.5.B66 2007
617.4'410592--dc22 2006015413

Preface

Molecular steps involved in the hematopoietic stem cell (HSC) activation and self-renewal have the potential for clinical use in the future. Stem cells can be quantified and sorted successfully, but transduction, labeling, and in vivo tracking of stem cells are difficult techniques. The in vivo bioluminescence imaging technology described in *Bone Marrow and Stem Cell Transplantation*, is a promising tool for stem cell transplant research models. The advent of human leukocyte antigen (HLA) had a major impact on the success of clinical transplantation. With the application of new molecular techniques to the field of HLA typing, posttransplant outcome has improved even further. HLA is one of the most polymorphic regions of the human genome, and molecular techniques are required for low- and high-resolution typing at a four-digit level. For certain HLA classes and alleles, serologic typing is insufficient and techniques such as SSO, SSP and even sequencing have become mandatory. HLA or minor-HLA compatibility between donors and recipients are important predictors of graft-vs-host (GVH) and graft-vs-disease (GVD) interactions. The disparity between donor–recipient pairs in regard to non-HLA polymorphisms such as killer inhibitory receptor, cytokine, hormone receptor or noncoding region polymorphisms are becoming important. Such methods are becoming more widely applicable and their impact on transplant outcome is currently under evaluation. Although some of these disparities are not influential on transplant outcome, they are useful by enabling quantitative measurement and dynamic monitorization of chimerism. Dynamic follow-up of engraftment kinetics allows efficient timely infusions of donor lymphocyte or anticancer intervention. Another application of molecular techniques to the field of HSC transplantation (HSCT), is the quantification of posttransplant residual disease. These techniques constitute an efficient approach to determination of cure or success of transplantation. However, there are many limitations in regard to the number of disorders suitable for such analyses, and the sensitivity of these techniques may differ. Also, the evolving biology and genetics of these disorders may, in time, cause additional problems. Another evolving field related to HSCT is the regenerative potential of stem cells. Detection of a low frequency of donor type non-hematopoietic cells in tissues such as liver, skin, and gastrointestinal tract following blood or marrow transplantation is a challenging field. The recent trend of moving from gene expression to the protein expression profiles has resulted in application of proteomics to posttransplant follow-up of patients. These premature results are promising for selection of candidate molecules in early and sensitive monitorization of GVHD.

As summarized in *Bone Marrow and Stem Cell Transplantation*, molecular techniques have contributed to the field of HSCT, and are becoming, if not already, routine applications.

Meral Beksac

Contents

Contributors

MERAL BEKSAC • *Department of Hematology, School of Medicine, Ankara University, Ankara, Turkey*
TERESA V. BOWMAN • *Center for Cell and Gene Therapy. Baylor College of Medicine, Houston, TX*
JULIA BRENMOEHL • *Department of Internal Medicine I, University Medical Centre of Regensburg, Regensburg, Germany*
MATTEO G. CARRABBA • *Division of Hematology, Istituto Nazionale Tumori,University of Milano, Milan, Italy*
CHRISTOPHER H. CONTAG • *Departments of Pediatrics, Microbiology and Immunology, and Radiology, Stanford University School of Medicine, Stanford, CA*
PAOLO CORRADINI • *Deparment of Hematology, Bone Marrow Transplantation Unit, Istituto Nazionale per lo Studio e la Cura dei Tumori, Università degli Studi di Milano, Milan, Italy*
KLARA DALVA • *Department of Hematology, Tissue Typing Laboratories, Ibni Sina Hospital, Ankara University, Ankara, Turkey*
M. TEVFIK DORAK • *Sir James Spence Institute, School of Clinical Medical Sciences (Child Health), University of Newcastle, UK*
JENNIFER DUDA • *Program in Molecular Imaging at Stanford (MIPS), Stanford University School of Medicine, Stanford, CA*
AHMET H. ELMAAGACLI • *Department of Bone Marrow Transplantation, University Hospital of Essen, Essen, Germany*
LUCIA FARINA • *Division of Hematology, Istituto Nazionale Tumori, University of Milano, Milan, Italy*
MARGARET A. GOODELL • *Center for Cell and Gene Therapy, Houston, TX*
ELS GOULMY • *Department of Immunohaematology and Blood Transfusion, Leiden University Medical Center, Leiden, The Netherlands*
ERNST HOLLER • *Division of Haematology/Oncology, University Medical Centre of Regensburg, Regensburg, Germany*
BERTHOLD HOPPE • *Institute of Transfusion Medicine, Charité – Universitätsklinikum Berlin, Berlin, Germany*
MOBIN KARIMI • *Stanford University School of Medicine, Stanford, CA*
ISINSU KUZU • *Department of Pathology, School of Medicine, Ankara University, Ankara, Turkey*

THOMAS LION • *Children´s Cancer Research Institute, St. Anna Children's Hospital, Vienna, Austria*
ANN-MARGARET LITTLE • *Histocompatibility Laboratories, The Anthony Nolan Trust, Royal Free Hospital, London, UK*
AKIL A. MERCHANT • *Center for Cell and Gene Therapy, Baylor College of Medicine, Houston, TX*
HARALD MISCHAK • *Departments of Hematology, Hemostasis, and Oncology, Hannover Medical School, Hannover, Germany*
PETE G. MIDDLETON • *Haematological Sciences, School of Clinical and Laboratory Sciences Medical SchoolUniversity of Newcastle, Newcastle Upon Tyne, UK*
ROBERT S. NEGRIN • *Department of Medicine, Stanford University, Stanford, CA*
ABDULGABAR SALAMA • *Universitätsklinikum Charité, Institut für Transfusionsmedizin Medizinische Fakultät der Humboldt-Universität zu Berlin, Berlin, Germany*
ERIC SPIERINGS • *Department of Immunohematology and Blood Transfusion, Leiden University Medical Center, Leiden, The Netherlands*
GERHARD ROGLER • *Department of Internal Medicine I, University Medical Centre of Regensburg, Regensburg, Germany*
EVA M. WEISSINGER • *Departments of Hematology, Hemostasis, and Oncology, Hannover Medical School, Hannover, Germany*

1

Molecular Profiling of Hematopoietic Stem Cells

Teresa V. Bowman, Akil A. Merchant, and Margaret A. Goodell

Summary

Gene expression profiling using microarrays is a powerful method for studying the biology of hematopoietic stem cells (HSCs). Here, we present methods for activating HSCs with the chemotherapeutic drug 5-Fluorouracil, isolating HSCs from whole bone marrow, and performing microarray analysis. We also discuss quality control criteria for identifying good arrays and bioinformatics strategies for analyzing them. Using these methods, we have characterized the gene expression signatures of HSC quiescence and proliferation and have constructed a molecular model of HSC activation and self-renewal.

Key Words: Hematopoietic stem cells; microarray; 5-Fluorouracil; gene expression.

1. Introduction

Murine hematopoietic stem cells (HSCs) are the best described adult stem cell population at a phenotypic and functional level. HSCs have two definitive characteristics: extensive self-renewal capacity and multipotentiality. Experimental evidence from mouse knockout and retroviral overexpression studies illuminated the importance of genetic components in regulating the decision between self-renewal and differentiation. Recognizing this importance, several investigators have taken genomic-based approaches to uncover the complex genetic regulation of HSCs using global expression profiling.

Despite advances in identifying and isolating functional HSCs, understanding the most coveted property of HSCs, the ability to faithfully self-renew without losing any potential remains elusive. In vivo HSCs appear to self-renew indefinitely, but once removed and cultured in vitro, their self-renewal capacity quickly diminishes. In vivo, HSCs exist in a state of quiescence, with around 1–3% in cycle and approx 98% in G0 ***(1–5)***. This was initially demonstrated experimentally by the administration of cytotoxic drugs such as the pyrimidine

From: *Methods in Molecular Medicine, vol. 134: Bone Marrow and Stem Cell Transplantation*
Edited by: M. Beksac © Humana Press Inc., Totowa, NJ

analog 5-fluorouracil (5FU) ***(6,7)*** and hydroxyurea (HU) ***(8,9)***. These drugs, which are lethal to proliferating cells, kill cycling hematopoietic progenitors but spare the rare HSCs thus proving that HSCs are indeed quiescent in vivo ***(1–5)***.

Treatment with cytotoxic drugs spares quiescent HSCs, but in doing so stimulates them into an active proliferative state. A single injection of 5FU kills most hematopoietic cells, and brings the remaining HSCs into cycle to repopulate the depleted bone marrow ***(6–8)***. HSC proliferation proceeds in a time-dependent manner, peaking 5–6 d after treatment with approx 20% of HSCs in cycle, before returning to normal, starting around day 10 **(Fig. 1A)** ***(8)***.

Advances in gene expression technology over the past decade have allowed scientists to generate global transcriptional profiles consisting of thousands of genes. Traditionally, differential gene expression could only be determined on a single-gene basis using techniques like reverse-transcription (RT)-PCR ***(9)*** and Northern blotting ***(10)***. Now, thousands of genes can be assessed simultaneously; therefore, increasing our knowledge of the transcriptome while also unveiling the complexity of gene regulation within a cell.

Several methods have been designed to study global gene expression. These technologies can be divided into two major categories: open and closed. Open systems are generally sequence-based systems requiring no *a priori* knowledge of the transcriptome, thus giving the investigator the potential for novel gene discovery. Closed systems are generally microarray-based systems assessing thousands of characterized and uncharacterized genes at once. This type of analysis allows for assessment of gene expression of the same set of transcripts for various cell types under every type of condition. These global gene expression procedures include differential display (DD)-PCR ***(11)***, representational difference analysis/suppression subtraction hydridization (RDA/SSH]) ***(12,13)***, serial analysis of gene expression (SAGE) ***(14,15***, expressed sequence tag (EST) ***(16)***, and nucleic acid array technology ***(17–19)***.

Recently, our lab and others have published methodologies to understand the genetic regulation underlying HSC proliferation ***(20,21)***. Using microarray technology, we compared populations of quiescent and proliferating HSCs,

Fig. 1. *(opposite page)* Hematopoietic stem cell (HSC) activation and purification. **(A)** Graphic depicting the changes in bone marrow cellularity and number of HSC in cell cycle following 5-fluorouracil (5FU) treatment. (Adapted from **refs.** ***7,8***.) **(B)** SP profiles of adult HSCs and 5FU-HSCs 6 d post 5FU treatment (marrow was enriched for Sca-1 expressing cells with magnetic beads prior to analysis). Arrows point to analysis in SP cells of Sca-1 and lineage marker expression showing greater than 97% homogeneity for Sca-1 and lineage expression. *(Continued on next page)*

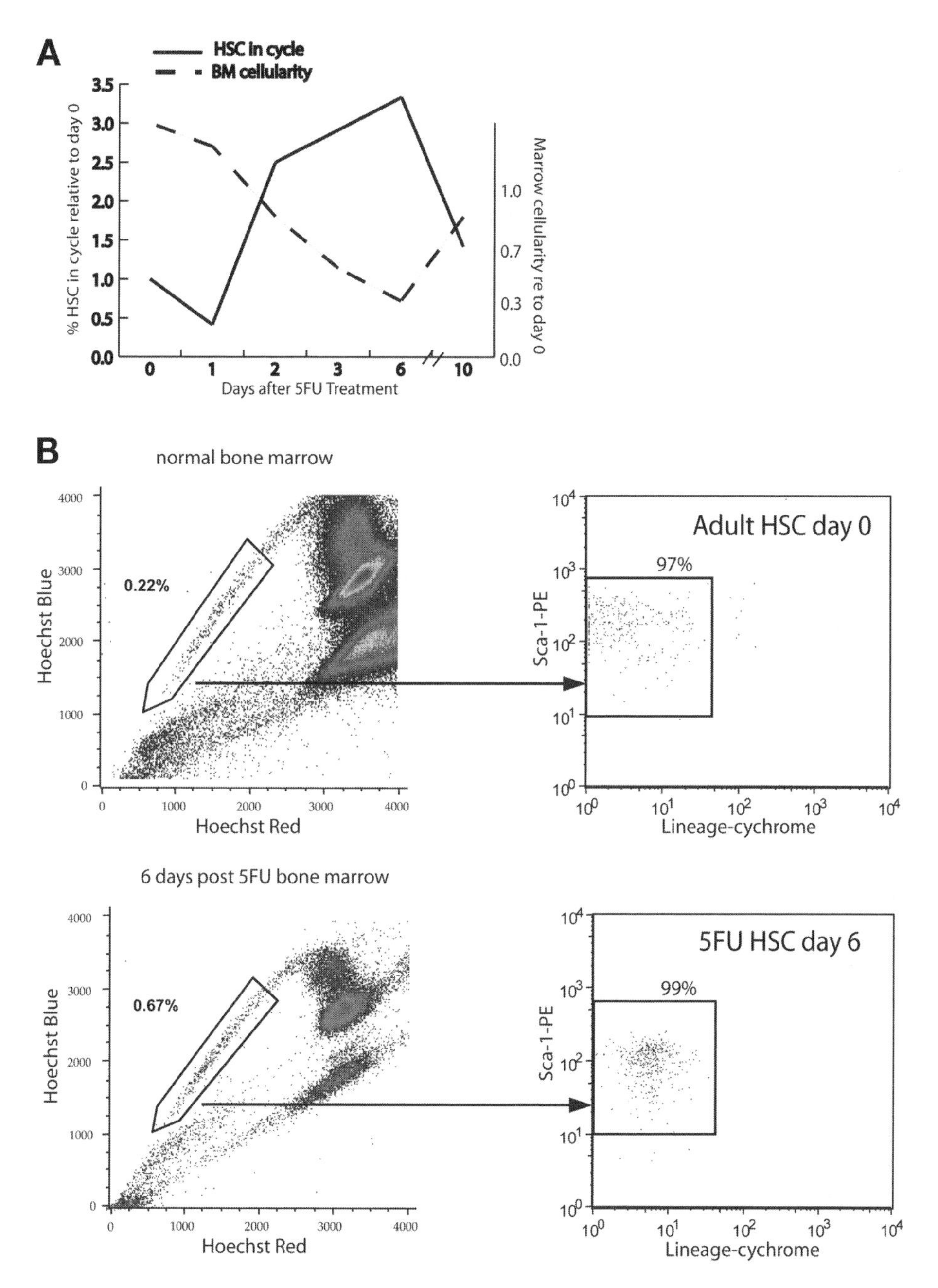

Fig. 1. *(Continued from previous page)* For analysis of adult HSCs on day 0, the lineage markers used were Mac1, CD4, CD8, B220, Gr1, and Ter 119. For analysis of 5FU-HSCs on day 6, all of the above markers were used except for Mac1, because of its low level expression on HSCs after 5FU treatment.

characterized the gene expression signatures of HSC quiescence and proliferation, and constructed a molecular model of HSC activation and self-renewal. This model has given us a framework in which to identify factors important for intiating, maintaining, and terminating the stem cell proliferative response. Many of these factors are likely to not only regulate normal HSC function but be involved in hematopoietic malignancies. Understanding this function promises to give us new insight into cancer biology and potentially identify new therapeutic targets.

2. Materials

2.1. Activation of Murine Hematopoietic Stem Cells With 5FU

1. 5FU (Sigma, cat. no. F6627)
2. 1X Dulbecco's phosphate-buffered saline (PBS), no calcium, no magnesium (Invitrogen, cat. no. 14190-144).
3. 5FU stock solution (50 mg/mL): resuspend 1 g of 5FU in 20 mL of 1X PBS. 5FU goes into solution slowly. Heat the mixture to 37°C and vortex often.
4. 5FU working solution (10 mg/mL): dilute 5FU stock solution 1:5 in 1X PBS to make a 10 mg/mL working solution. For a dose of 150 mg/kg body weight, each mouse (of approx 20 g) should receive 300 μL of this solution. Make a fresh batch of working solution for each set of injections.

2.2. HSC Isolation and Purification

1. C57Bl/6 mice (Jackson laboratories; www.jax.org).
2. Dissecting tools: scissors and forceps (Drummond Scientific tools are recommended).
3. Syringes: 10 or 20 mL.
4. Needles: 18.5 and 27.5 gauge.
5. Tissue culture plates.
6. Hank's balanced salt solution (HBSS; Invitrogen, cat. no. 14170-112).
7. Dulbecco's modified Eagle's medium (DMEM; Invitrogen, cat. no. 119650-092).
8. Fetal bovine serum (FBS; Invitrogen, cat. no. 26140-079).
9. 1 *M* HEPES (Invitrogen, cat. no. 15630-106).
10. HBSS+: HBSS + 2% FBS + 1% 1 *M* HEPES.
11. DMEM+: DMEM + 2% FBS + 1% 1 *M* HEPES.
12. Cell strainers: 40 and 70 μm.
13. Hemacytometer.
14. Centrifuge.
15. Hoechst 33342 (bis-benzimide, Sigma, cat. no. B2261).
16. Hoechst stock solution (1 mg/mL, 200X): resuspend Hoechst powder in water to a final concentration of 1 mg/mL, filter-sterilize, and freeze aliquots at –20°C.
17. Circulating water bath at 37°C.
18. Antibodies: c-Kit-FITC (clone 2B8), Sca-1-PE (clone E13-161.7), GR1-FITC (clone RB6-8C5), GR1-PE (clone RB6-8C5) from BDPharmingen; CD4-PE-Cy5

(clone L3T4), CD8-PE-Cy5 (clone 53-6.7), B220-PE-Cy5 (RA3-6B2), Ter119-PE-Cy5 (clone Ter119), GR1-PE-Cy5 (clone RB6-8C5), Mac1/CD11b-PE-Cy5 (clone M1/70) from Ebiosciences.

19. Propidium iodide (PI; Sigma, cat. no. P4170).
20. PI stock solution (200 μg/mL, 100X): resuspend PI powder in water to a final concentration of 200 μg/mL, filter-sterilize, and freeze aliquots at –20°C.
21. Fluorescence-activated cell sorter (FACS) equipped with proper lasers and filters to excite and detect Hoechst 33342 ***(22)***.

2.3. RNA Isolation

1. RNAqueous RNA Isolation Kit (Ambion, cat. no. 1912).
2. 100% Ethanol.
3. 70% Ethanol made from 100% ethanol.
4. Heat block.
5. Nonstick RNAse-free 1.5-mL tubes (Ambion, cat. no. 12450).
6. DNaseI, amplification grade (Invitrogen, cat. no. 18068-015).
7. Phenol:chloroform:isoamyl alcohol (25:24:1) (Sigma, cat. no. P3803).
8. Phase Lock Gel: PLG-Light, 0.5-mL tubes (Eppendorf, cat. no. 0032005.004).
9. Microcentrifuge.
10. Ultrapure glycogen (Invitrogen, cat. no. 10814-010).
11. 3 *M* Sodium acetate (Ambion, cat. no. 9740).
12. Nuclease-free water (Ambion, cat. no. 9937).

2.4. RNA Amplification

1. MessageAmp amplified RNA (aRNA) Amplification Kit (Ambion, cat. no. 1750) or MessageAmp II aRNA Amplification Kit (Ambion, cat. no. 1751).
2. Nuclease-free water (Ambion, cat. no. 9937).
3. Thermal cycler with heated lid.
4. Hybridization oven.
5. Microcentrifuge with vacuum manifold.
6. Biotinylated nucleotides: biotin-11-CTP (ENZO, P/N 42818) and biotin-16-UTP (ENZO, P/N 42814).
7. 5X Fragmentation buffer: 4 mL 1 *M* Tris acetate pH 8.1, 0.64 g magnesium acetate, 0.98 g potassium acetate, and nuclease-free water to a final volume of 20 mL.
8. Poly-A RNA Control Kit (Affymetrix, cat. no. P/N 900433).

2.5. Microarray Hybridization

Note: Subheading 2.5., items 1–4 and **Subheading 2.6., item 1** are generally supplied and run by an institute's microarray core facility (free download from: http://www.affymetrix.com/support/technical/product_updates/gcos_download.affx).

1. Affymetrix microarrays: (Mouse MOE430 2.0).
2. Hybridization Oven 640.

3. GeneChip Fluidics Station 450.
4. GeneChip Scanner 3000.

2.6. Low-Level Analysis: Microarray Normalization, Expression Measure Determination, and Quality Control

1. GCOS software.
2. R. Download latest version from http://cran.r-project.org R is open source statistical analysis software.
3. Bioconductor packages. Download latest version from: www.bioconductor.org. Bioconductor is an open source and open development software project to provide tools for the analysis and comprehension of genomic data.

2.7. Higher-Level Analysis

1. Bioconductor packages for gene list analysis: *simpleaffy*-generates QC information and allow for very basic analysis; *genefilter*-tools to sequentially filter genes to generate gene lists; *multitest*-tools for multiple testing procedures for controlling family-wise error rate (FWER) and false discovery rate (FDR); *Siggene*-tool to identify differentially expressed genes and estimate FDR with both significance analysis of microarrays (SAM) and empirical Bayes analyses of microarrays.
2. Scripts for regression analysis of time course ***(20)***. Gene ontology (GO) tools: Ontology Traverser (franklin.imgen.bcm.tmc.edu/rho-old/services/OntologyTraverser/) ***(23)***; Bioconductor packages: GOstats.html, goTools, ontoTools (www.bioconductor.org); EASEv2.0 (david.niaid.nih.gov/david/ease.htm).

3. Methods

3.1. Activation of Murine Hematopoietic S tem Cells With the Cytotoxic Drug 5FU

1. To administer the 5FU, inject mice intravenously with a single dose of 150 mg/kg body weight. An adult mouse 8–12 wk old weighs approx 20 g, thus each mouse receives approx 3 mg of 5FU. Working solution should be diluted fresh to a concentration of 10 mg/mL. Each mouse should be injected with 300 μL of working solution.
2. Effectiveness of 5FU cytotoxicity can be monitored by changes in peripheral blood cell counts and bone marrow cellularity **(Fig. 1A)**.

3.2. Isolation and Purification of Murine Hematopoietic Stem Cells (see **Note 1**)

1. Prewarm DMEM+ at 37°C.
2. Bone marrow cells are isolated from the tibia and femur of adult mice (8–12 wk old) before or after 5FU injection. Femurs and tibias are removed from euthanized animals.

3. After bone removal, bone marrow is flushed from the bones using a syringe and a 27.5-gage needle filled with HBSS+.
4. The flushed marrow is then triturated into a single cell suspension using a new syringe and an 18.5-gage needle.
5. Following trituration, the single cell suspension is filtered through a 70-μm cell strainer to remove any debris and large cell clumps.
6. Count the nucleated cells using a hemacytometer. Accuracy in counting is important to ensure proper Hoechst staining. For C57Bl/6 mice of this age, approx 5×10^7 nucleated cells are isolated using this flushing method.
7. Bone marrow cell suspension is pelleted by centrifugation at 1500*g* for 8 min. Pour off the supernatant.
8. For Hoechst staining, the cell pellet is resuspended in prewarmed DMEM+ at a final concentration of 10^6 cells/mL. Hoechst 33342 dye is added to a final concentration of 5 μg/mL (use a 1:200 dilution of the Hoechst stock solution.)
9. Bone marrow cells are incubated in a circulating water bath set at 37°C for 90 min. Timing is crucial.
10. All steps after Hoechst staining must be performed at 4°C (*see* **Note 2**). Bone marrow cells are pelleted by centrifugation at 1500*g* for 8 min at 4°C. Pour off the supernatant.
11. For antibody staining, the cell pellet is resuspended in ice-cold HBSS+ medium at a final cell concentration of 10^8 cells/mL. Four aliquots of 200,000 cells each are removed and stained as compensation controls for the flow cytometry. The details of antibody staining of the controls will be discussed later.
12. Magnetic enrichment of Sca-1- or c-Kit-positive cells can be performed at this step (*see* **Note 3**).
13. The remaining cells are stained with the following antibodies each at a 1:100 dilution: c-Kit conjugated to fluorescein isothiocyanate (FITC), Sca-1 conjugated to PE, and lineage markers conjugated to *R*-phycoerythrin (PE)-Cy5 (CD4, CD8, B220, Ter119, Gr1, and Mac1/CD11b). For bone marrow cells isolated after 5FU treatment, c-Kit and Mac1/CD11b were excluded from the cocktail as they change their expression pattern after HSC activation (**Fig. 1B**).
14. Cells are incubated on ice for 15 min to allow for antibody–antigen interaction. Cells are pelletted at 1500*g* for 8 min. Pour off the supernatant.
15. Resuspend the cell pellet in ice-cold HBSS+ supplemented with 2 μg/mL PI at a final cell concentration of $2–4 \times 10^7$ cells/mL. Filter cells through a 40-μm cell strainer before running samples on a FACS.
16. Prepare compensation controls by staining each control singly with a common cell surface marker conjugated to each of the fluorochromes used. In the above experiment this would include FITC, PE, and PE-Cy5. Controls: (1) Hoechst-stained only, (2) Hoechst stained + anti-Gr1 conjugated to FITC, (3) Hoechst stained + anti-Gr1 conjugated to PE, and (4) Hoechst stained + anti-Gr1 conjugated to PE-Cy5. Each control was stained with antibodies on ice for 15 min. Controls are pelleted by centrifugation at 1500*g* for 8 min (*see* **Note 4**). Pour off supernatant.

17. Resuspend controls in 500 μL of HBSS+ supplemented with 2 μg/mL PI.
18. For detection of the Hoechst-stained cells, an ultraviolet (UV) laser at 350 nm is needed to excite the dye. The emitted fluorescence is measured at two wavelengths: 450 nm (Hoechst Blue) and 675 nm (Hoechst Red) *(22)*.
19. After sorting HSCs, perform a purity check on the sorted cells. For experiments on gene expression, it is critical to have an extremely pure cell population. For HSCs, it is important to have a ≥95% purity for side population and cell surface markers. It is also important to exclude red blood cells because these cells are full of highly abundant transcripts such as hemoglobin that could skew the final expression results.

3.3. RNA Isolation

1. Sorted HSCs are pelleted by centrifugation at 1500*g* for 8 min. Remove supernatant carefully so as not to disturb the small pellet. It is better to leave a couple of microliters of liquid behind than to lose any of your cells.
2. Resuspend HSCs in 200 μL RNAqueous lysis buffer. Triturate pellet well to ensure complete cell lysis. Let the lysis stand at room temperature for 5 min. Sample may be frozen at this step at –80°C for several months.
3. To isolate, follow the instructions from the RNAqueous RNA isolation kit. One modification to the protocol is listed: to elute the RNA, add 35 μL of preheated elution solution to the column, allow to incubate for 1 min, and then centrifuge for 1 min. Repeat the elution step again.
4. Following elution of RNA, immediately treat with DNaseI to remove any residual genomic DNA. To the isolated RNA, add 10 μL 10X DNaseI buffer, 10 μL DNaseI, 2 μL RNAse inhibitor, and 8 μL nuclease-free water. The best results for the DNA digestion were observed when the reaction was performed at 37°C for 30 min. Note that this differs from the manufacturer's suggested protocol.
5. Stop the reaction with addition of 10 μL 25 m*M* EDTA.
6. For further purification of RNA, follow DNaseI treatment with a phenol:chloroform extraction. Add 110 μL of phenol:chloroform:isoamyl alcohol to the RNA/DNaseI solution and mix the solution by vigorous shaking.
7. To separate the aqueous and organic phases, add the mixture to a 0.5-mL phase-lock gel tube and mix by shaking. Centrifuge the solution at 10,000*g* for 5 min. Separation should be complete; the aqueous phase containing the RNA will be on the top separated from the organic phase by the phase-lock gel.
8. Precipitate the RNA with 30 μL 3 *M* sodium acetate + 300 μL 100% ethanol + 1 μL glycogen (to aid visualization of pellet—*see* **Note 5**)
9. Allow RNA to precipitate at –80°C for 30 min to overnight. Centrifuge the RNA at 10,000*g* at 4°C for 15 min.
10. Carefully remove the supernatant being sure not to disturb the pellet. Wash the pellet by adding 1 mL 70% ethanol followed by centrifugation at 14,000 rpm at 4°C for 10 min.
11. Carefully remove the supernatant being sure not to disturb the pellet. Using a 20-μL pipet tip, remove any residual alcohol, as this might inhibit further reactions.

12. Allow the sample to briefly air dry. Resuspend the pellet in 11 μL of nuclease-free water. (Resuspend 12 μL if performing analysis on NanoDrop as described in **step 13**.)
13. Determine concentration of RNA by spectrophotometry. We recommend using a NanoDrop-1000, because only a small amount of material (1 μL) is needed for accurate concentration determination. Using the previously listed procedure, 4–5 pg of total RNA can be isolated per murine HSC.

3.4. RNA Amplification

For Affymetrix microarrays, 20 μg of antisense RNA is needed for hybridization. Because only nanogram quantities of RNA are isolated from HSCs, RNA must be amplified to produce enough material for the hybridization. The MessageAmp aRNA synthesis kit is used to amplify RNA. Affymetrix also offers an alternate protocol that uses the same strategy, but does not supply all the components as a single kit. If desired, Poly-A RNA spike in controls can be added to the RNA sample prior to amplification.

Perform RNA amplification according to manufacturer's instructions. Any modifications will be listed as follows:

1. First round amplification–first strand synthesis: the first step in the RNA amplification procedure is to convert the RNA into complementary DNA (cDNA). Incubate the reaction at 42°C in a thermal cycler with a heated lid for 2 h.
2. First round amplification–second strand synthesis: incubate the mixture at 16°C in a thermal cycler for 2 h.
3. First round amplification—cDNA purification.
4. First round amplification—in vitro transcription of T7 cDNA: the first in vitro transcription reaction is performed to amplify the starting material. The aRNA produced from this reaction will not be hybridized on a microarray, thus modified nucleotides will not be used. At the end of the first in vitro transcription, we generally have generated 0.5–2 μg of aRNA. Incubate the reaction at 37°C for 16–24 h in a dry heat hybridization oven.
5. First round amplification—aRNA purification.
6. Second round amplification—first strand synthesis: we use 500 ng of aRNA for the second round, or if yield from the first round is less than 500 ng, all of the first round sample is used. RNA should be at 50 ng/μL and may need to be concentrated using a vacuum manifold. Do not use heat >42°C during concentration, as higher temperatures could induce RNA degradation. Incubate the first strand reaction at 42°C in a thermal cycler with a heated lid for 2 h.
7. Second round amplification—second strand synthesis: incubate the mixture at 16°C in a thermal cycler for 2 h.
8. Second round amplification—cDNA purification.
9. Second round amplification—in vitro transcription: the product from this reaction will be hybridized on an Affymetrix microarray, thus biotin-modified nucleotides will be used. At the end of the second in vitro transcription, we typi-

cally generated 50–100 mg of aRNA, ≥10,000-fold amplification. Incubate the reaction at 37°C in a dry heat hybridization oven for 16–24 h.

10. Second round amplification—aRNA purification.
11. Second round amplification—aRNA fragmentation.
    a. Determine concentration of aRNA using spectrophotometry.
    b. Bring the RNA to a concentration of ≥0.6 μg/μL. Fragment the material by combining 20 μg of aRNA, 8 μL of 5X fragmentation buffer, and nuclease-free water to a final volume of 40 μL.
    c. Incubate the mixture at 94°C for 35 min in a thermal cycler with a heated lid to allow for autocatalysis of the RNA into smaller fragments of an average length of 50 bp.
    d. Stop the reaction by placing the tube on ice (4°C).
    e. Amplified and fragmented RNA can be examined on a 2% agarose gel or using an Agilent Bioanalyzer. aRNA (two rounds) should appear as a smear with the majority of RNA around 500 bp (as visualized with a standard DNA ladder). Fragmented RNA should appear between 25 and 50 bp (DNA ladder) (**Fig. 2A**).
    f. aRNA is now ready for hybridization.

3.5. Microarray Hybridization

1. Fragmented biotinylated aRNA can be hybridized to Affymetrix microarrays according to standard protocols. Basic steps include hybridization, washing, staining, washing, and scanning and are typically performed by a core facility. The procedure takes up to 2 d.
2. Raw image (DAT) and intensity (CEL) files are generated using GCOS software (www.affymetrix.com).

3.6. Low-Level Analysis: Microarray Normalization, Expression Measure Determination, and Quality Control

1. Affymetrix provides a primer for data analysis than can be accessed at http://www.affymetrix.com/Auth/support/downloads/manuals/data_analysis_fundamentals_manual.pdf.
2. Image files should be visually inspected for large defects in either manufacturing or resulting from the hybridization procedure. **Figure 3A** depicts a large spot in the center of the chip that is likely a hybridization artifact (*see* **Note 6**).
3. Image files (.DAT) are gridded to determine the location of each probe spot. This gridding should be manually checked to ensure proper calculation of probe intensity values (requires GCOS or MAS 5.0 software from Affymetrix).
4. The intensity files (.CEL) generate many parameters that can be used to assess quality of the experiment. Investigators should check scale factor, noise (Raw Q), average Background, and 3′/5′ ratio of control genes (GAPDH and β-actin) *(24)* (**Table 1**).
5. Affymetrix provides Poly-A RNA controls that can be spiked into the RNA to evaluate the amplification and labeling process. Poly-A controls consist of the

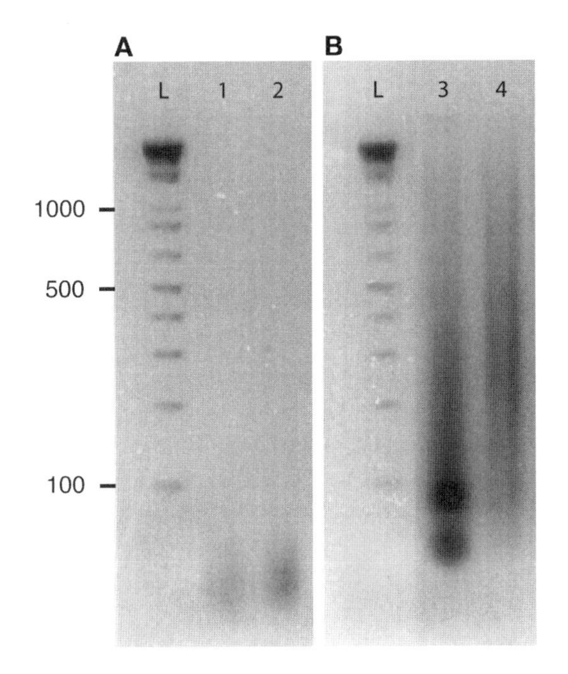

Fig. 2. Amplified RNA (aRNA) before and after fragmentation. **(A)** Fragmented aRNA. Lanes 1 and 2 show fragmented aRNA at approx 50 bp. **(B)** RNA amplified with two rounds of linear amplification, before fragmentation. The RNA in lane 4 shows good amplification with the smear centered at approx 500 bp. The RNA in lane 3 was poorly amplified, with a predominance of smaller transcripts. Chips hybridized with this RNA showed high scale factors, low presence calls, and high 3′/5′ ratios of house keeping genes (*see* **Table 1** for details). (Note: a standard DNA ladder is being used for convenience, but may not accurately reflect the exact RNA sizes.)

B. subtilis genes *lys, phe, thr*, and *dap,* and should all appear "present" on the chip with increasing signal in their listed order ***(24)***.

6. Intensity files are the direct input that is utilized as the data file in the Bioconductor packages for Affymetrix chip analysis (*see* **Subheading 2.**).
7. Normalization, background correction, and model-based expression values can be calculated using gcrma from Bioconductor. GC-RMA is a method specifically designed to improve the preprocessing of Affymetrix microarray data. Hundreds of thousands of probes exist on a single microarray. High levels of sequence similarity are unavoidable. Cross-hybridization will occur, which results in an overestimation of the "true" or gene-specific signal of a given probe. The GC-RMA method performs a background adjustment based on sequence information of each probe to minimize the effects of cross-hybridization. The result is a better estimate of the "true" amount of starting mRNA in a sample.

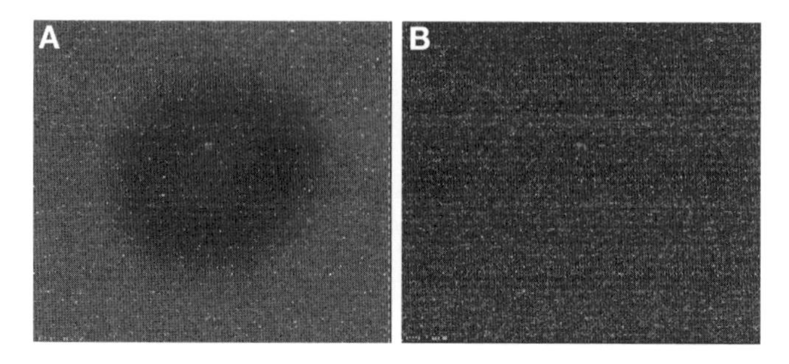

Fig. 3. Hybridization artifact. **(A)** Bad chip. Note the large spot in the center of the chip with the area of high background surrounding. This may have been owing to loading inadequate amounts of hybridization cocktail or a small leak from the chip. **(B)** Normal chip provided for comparison.

3.7. Higher-Level Analysis

1. Once quality control, low-level processing, and normalization are complete, higher-level analysis can begin. The type of analysis used is integral to the overall experimental design, has an important impact on number of replicate chips and types of control chips needed, and should be among the first considerations when beginning microarray experimentation. Analysis strategies are limited only by the creativity of the investigators; therefore, a detailed catalog of such strategies is not possible. We describe as follows, in general terms, some of the most common strategies, which we hope will serve as a starting point for further analysis.
2. The simplest experimental design is a pairwise comparison between two groups—for example, HSC vs non-HSC or mutant HSC vs wild-type HSC. The Bioconductor packages *simpleaffy*, *siggene*, or *genefilter* can be used for making a pairwise comparison. *Siggene* identifies differentially expressed genes and estimates false discovery rates. *Genefilter* filters genes using a variety of filtering functions such as coefficient of variation of expression measures, analysis of variance (ANOVA) *p*-value, and Cox model *p*-values.
3. Time course experiments involve measuring gene expression at several time points during a process, and can be a very powerful strategy for understanding complicated biology. Time courses can be developmental (different embryonic stages, for example), center around a biological process (such as cell cycle), or follow a tissue response to a drug or other treatment. For our time course experiment of HSC activation, we analyzed the 5FU time-dependent expression profiles for each gene by regressing the normalized expression values onto a polynomial basis using least squares regression. ANOVA was performed on the coefficients of regression to identify genes with significant time patterns ($p < 0.05$). The smooth curve fit approach assumed that the expression trajectory for each gene

Table 1
Affymetrix Quality Control Parameters

Parameter	Expected value	Comments
Visual artifact	None	
Proper gridding	Yes	
Scale factor	Maximum threefold difference between samples.	Varies. For our experiments, we discarded chips with scale factor greater than 10.
Raw Q	Varies. Should be similar between replicates.	Electrical noise from the scanner is the biggest contributor. Affymetrix does not suggest normal values, but they should be similar for all chips in an experiment and nearly identical between technical replicates.
Average background	20–100	
Percent present	Varies according to tissue. Typically between 30 and 55%.	Bad arrays will have a significantly lower percent present compared to other chips in the experiment.
3′/5′ Ratio of β-actin and GAPDH	Ideally close to 1, with a maximum of 3.	When using two rounds of amplification, use the 3′/middle ratio. When amplifying very small amounts of starting RNA, we tolerate higher ratios of up to 10.

followed a continuous time pattern. R scripts for this analysis method are available as supplemental data in **ref. *20***.

4. Making sense of gene lists: GO (www.geneontology.org) can be used to determine biologically meaningful patterns from data once gene lists are generated. GO is a controlled vocabulary that describes gene functions in their cellular context and is arranged in a quasi-hierarchical structure from more general to more specific. Because the vocabulary of annotations is fixed, it allows for functional comparisons of mutually exclusive gene lists. The Shaw lab provides a web-based tool for conducting GO-based analysis and can be accessed at franklin.imgen.bcm.tmc.edu/rho-old/services/OntologyTraverser/.

4. Notes

1. For a detailed description of the isolation and purification of murine HSCs, *see* **ref. *22***.
2. SP cells are distinguished from other cells in the bone marrow according to dye efflux by ABC transporter pumping. This process is ATP-dependent and temperature-dependent. The process can be halted or slowed on ice (4°C).
3. Magnetic enrichment can be performed to aid the purity of cell sorting. Miltenyi Biotec has many schemes to enrich or deplete murine and human blood and bone marrow.
4. Compensation controls should have clear positive and negative cell populations. These controls are used to set up the instrument properly to clearly distinguish fluorochromes with overlapping emission spectra.
5. Glycogen, which is used to aid in visualizing the RNA pellet, is an inert molecule and should not affect later steps.
6. For an interesting gallery of images from abnormal chips, *see* http://stat-www.berkeley.edu/users/bolstad/PLMImageGallery/index.html.

Acknowledgments

MAG is a scholar of the Leukemia and Lymphoma Society. AAM is Molecular Medicine Scholar. We thank Stuart Chambers for critical reading of the manuscript and Chad Shaw for guidance with statistical analysis.

References

1. Goodell, M. A., Brose, K., Paradis G.,, Conner, A. S., and Mulligan, R. C. (1996) Isolation and functional properties of murine hematopoietic stem cells that are replicating in vivo. *J. Exp. Med.* **183,** 1797–1806.
2. Bradford, G. B., Williams, B., Rossi, R., and Bertoncello, I. (1997) Quiescence, cycling, and turnover in the primitive hematopoietic stem cell compartment. *Exp. Hematol.* **25,** 445–453.
3. Morrison, S. J. and Weissman, I. L. (1994) The long-term repopulating subset of hematopoietic stem cells is deterministic and isolatable by phenotype. *Immunity* **1,** 661–673.

4. Cheshier, S. H., Morrison, S. J., Liao, X., and Weissman, I. L. (1999) In vivo proliferation and cell cycle kinetics of long-term self-renewing hematopoietic stem cells. *Proc. Natl. Acad. Sci. USA* **96,** 3120–3125.
5. Arai, F., Hirao, A., Ohmura, M., et al. (2004) Tie2/angiopoietin-1 signaling regulates hematopoietic stem cell quiescence in the bone marrow niche. *Cell* **118,** 149–161.
6. Van Zant, G. (1984) Studies of hematopoietic stem cells spared by 5-fluorouracil. *J. Exp. Med.* **159,** 679–690.
7. Harrison, D. E. and Lerner, C. P. (1991) Most primitive hematopoietic stem cells are stimulated to cycle rapidly after treatment with 5-fluorouracil. *Blood* **78,** 1237–1240.
8. Randall, T. D. and Weissman, I. L. (1997) Phenotypic and functional changes induced at the clonal level in hematopoietic stem cells after 5-fluorouracil treatment. *Blood* **89,** 3596–3606.
9. Hart, C., Schochetman, G., Spira, T., et al. (1988) Direct detection of HIV RNA expression in seropositive subjects. *Lancet* **2,** 596–599.
10. Eikhom, T. S., Abraham, K. A., and Dowben, R. M. (1975) Ribosomal RNA metabolism in synchronized plasmacytoma cells. *Exp. Cell Res.* **91,** 301–309.
11. Liang, P. and Pardee, A. B. (1992) Differential display of eukaryotic messenger RNA by means of the polymerase chain reaction. *Science* **257,** 967–971.
12. Lisitsyn, N. and Wigler, M. (1993) Cloning the differences between two complex genomes. *Science* **259,** 946–951.
13. Diatchenko, L., Lau, Y. F., Campbell, A. P., et al. (1996) Suppression subtractive hybridization: a method for generating differentially regulated or tissue-specific cDNA probes and libraries. *Proc. Natl. Acad. Sci. USA* **93,** 6025–6030.
14. Velculescu, V. E., Zhang, L., Vogelstein, B., and Kinzler, K. W. (1995) Serial analysis of gene expression. *Science* **270,** 484–487.
15. Datson, N. A., van der Perk-de Jong, J., van den Berg, M. P., de Kloet, E. R., and Vreugdenhil, E. (1999) MicroSAGE: a modified procedure for serial analysis of gene expression in limited amounts of tissue. *Nucleic Acids Res.* **27,** 1300–1307.
16. Adams, M. D., Soares, M. B., Kerlavage, A. R., Fields, C., and Venter, J. C. (1993) Rapid cDNA sequencing (expressed sequence tags) from a directionally cloned human infant brain cDNA library. *Nat. Genet.* **4,** 373–380.
17. Lockhart, D. J., Dong, H., Byrne, M. C., et al. (1996) Expression monitoring by hybridization to high-density oligonucleotide arrays. *Nat. Biotechnol.* **14,** 1675–1680.
18. Pietu, G., Alibert, O., Guichard, V., et al. (1996) Novel gene transcripts preferentially expressed in human muscles revealed by quantitative hybridization of a high density cDNA array. *Genome Res.* **6,** 492–503.
19. Schena, M., Shalon, D., Davis, R. W., and Brown, P. O. (1995) Quantitative monitoring of gene expression patterns with a complementary DNA microarray. *Science* **270,** 467–470.
20. Venezia, T. A., Merchant, A. A., Ramos, C. A., et al. (2004) Molecular signatures of proliferation and quiescence in hematopoietic stem cells. *PLoS Biol.* **2,** e301.

21. Lambert, J. F., Liu, M., Colvin, G. A., et al. (2003) Marrow stem cells shift gene expression and engraftment phenotype with cell cycle transit. *J. Exp. Med.* **197,** 1563–1572.
22. Goodell, M. A., McKinney-Freeman, S., and Camargo, F. D. (2005) Isolation and characterization of side population cells. *Methods Mol. Biol.* **290,** 343–352.
23. Young, A., Whitehouse, N., Cho, J., and Shaw, C. (2005) OntologyTraverser: an R package for GO analysis. *Bioinformatics* **21,** 275–276.
24. Affymetrix. (2004) GeneChip Expression Analysis: Data Analysis Fundamentals.

2

Methods for Imaging Cell Fates in Hematopoiesis

Jennifer Duda, Mobin Karimi,
Robert S. Negrin, and Christopher H. Contag

Summary

Modern imaging technologies that allow for in vivo monitoring of cells in intact research subjects have opened up broad new areas of investigation. In the field of hematopoiesis and stem cell research, studies of cell trafficking involved in injury repair and hematopoietic engraftment have made great progress using these new tools. Multiple imaging modalities are available, each with its own advantages and disadvantages, depending on the specific application. For mouse models, clinically validated technologies such as magnetic resonance imaging (MRI) and positron emission tomography (PET) have been joined by optical imaging techniques such as in vivo bioluminescence imaging (BLI) and fluorescence imaging, and all have been used to monitor bone marrow and stem cells after transplantation into mice. Each modality requires that the cells of interest be marked with a distinct label that makes them uniquely visible using that technology. For each modality, there are several labels to choose from. Finally, multiple methods for applying these different labels are available. This chapter provides an overview of the imaging technologies and commonly used labels for each, as well as detailed protocols for gene delivery into hematopoietic cells for the purposes of applying these labels. The goal of this chapter is to provide adequate background information to allow the design and implementation of an experimental system for in vivo imaging in mice.

Key Words: Molecular imaging; hematopoiesis; bioluminescence; biophotonic imaging.

1. Introduction

In vivo imaging of labeled transplanted cells has made it possible to observe cell movement over time in the intact animal. Previous techniques have required sacrificing animals in order to analyze specific regions of interest for the presence of the rare labeled cells. Although these sorts of analyses certainly have yielded valuable information, their two major limitations are (1) each animal can

From: *Methods in Molecular Medicine, vol. 134: Bone Marrow and Stem Cell Transplantation*
Edited by: M. Beksac © Humana Press Inc., Totowa, NJ

only be analyzed at a single time point and (2) regions of interest must be preselected so potentially informative data on cell trafficking in other regions is lost. The sampling limitations and the lack of temporal information can introduce biases into the study. Therefore, a number of groups have developed noninvasive assays that can be performed repeatedly in a given animal. Use of such approaches in our group has revealed features of biological processes that were unexpected. These include an apparent tropism of *Listeria monocytogenes* for the lumen of the gall bladder—an otherwise sterile environment *(1)*, and the dynamic nature of hematopoietic reconstitution *(2)* during the earliest stages after stem cell transplantation (**Figs. 1** and **2**). In the stem cell studies it was observed that transplantation of a single hematopietic stem cell would result in a single focus of engraftment and expansion; however, some of these foci led to reconstitution and others did not (**Fig. 2**). These types of discoveries are only possible because of the ability to make longitudinal observations of the whole mouse after the transfer of labeled cells.

A number of powerful imaging techniques have been described and are now being used to understand when, where, and which cells move in many different mouse models of human biology and disease. Imaging has been applied to the study of injury repair, infectious processes, host response to infection, hematopoietic reconstitution, tumor vascularization, response to chemotherapy, and antitumor immune response. This chapter will focus on the labeling and imaging of murine bone marrow cells for the purpose of transplantation and analysis using in vivo imaging technologies. These types of studies can serve as a demonstration of how imaging can improve the study of biology and reveal nuances of complex biological processes.

1.1. Imaging Modalities

Several imaging modalities have been used for the investigation of stem cell fates and function and a number of useful tools have been developed for cell fate imaging by each of these *(3)*. Images of bone marrow and immune cell trafficking have been obtained by magnetic resonance imaging (MRI), positron emission tomography (PET), and single photon emission computed tomography (SPECT) *(4–9)*. Each of these is used clinically for diagnostic applications. In vivo bioluminescence (BLI) and fluorescence imaging may have their greatest application in the study of small animal models, and have been used to assess the trafficking patterns of cells in vivo *(1,2,6,10)*. There may be clinical applications of these tools, but their use in small animal models will be invaluable as they are applied to the study of stem cell biology. The availability of specialized MRI, PET, and SPECT scanners for small animal imaging has increased dramatically with improved performance and capabilities. Their increased use in animal models will likely lead to translational approaches, and

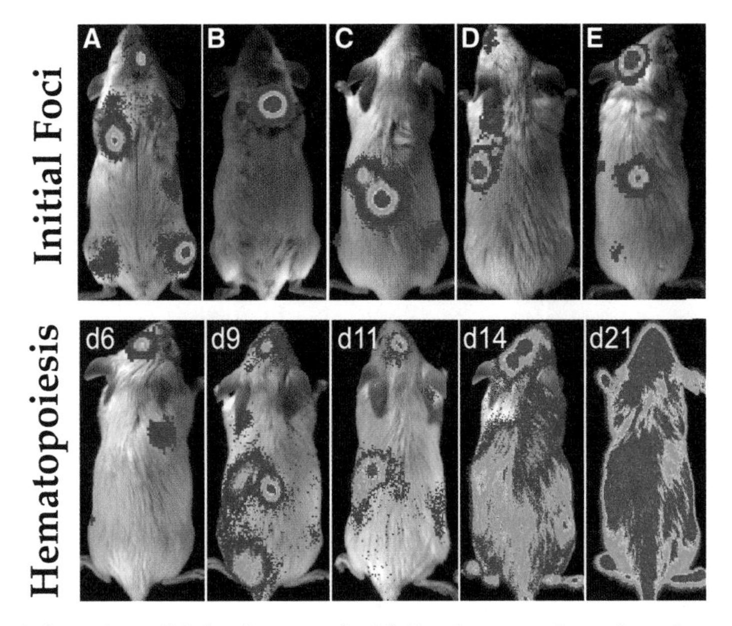

Fig. 1. Location of bioluminescent foci following transplantation of transgenic luc+ hematopoietic stem cells (HSCs). Foci were apparent in individual animals at anatomic sites corresponding to the location of the spleen, skull, vertebrae, femurs, and sternum (**A–E,** respectively) at 6–9 d after transfer. The patterns of engraftment were dynamic with formation and expansion or formation and loss of the bioluminescent foci. One recipient of 250 HSCs was monitored over time is shown (second row). In this animal, two initial foci were apparent on d 6. By d 9, one was no longer detectable and another remained at nearly the same intensity as on day 6. New foci were apparent on d 9, and then the intensity at these sites weakened or disappeared by d 11. Pseudocolored images reflecting optical signal intensity are overlaid upon a grayscale reference image of the animal. (Reprinted from **ref.** ***2,*** with permission.)

they hold great promise for advances in a number of fields where there are clinical needs. There are a number of optical imaging approaches that have been developed for sensing and imaging in the living body, including diffuse optical tomography (DOT), optical coherence tomography (OCT), and others, but only in vivo BLI and fluorescence have been used to study cell trafficking patterns. Despite the strengths of each of the available imaging modalities, the relatively greater accessibility, ease of use and throughput capabilities have contributed greatly to the widespread use of bioluminescent and fluorescent imaging for animal studies.

MRI uses magnetic fields and radio frequency pulses to induce and measure signals from hydrogen atoms in the body. The image resolution is excellent in

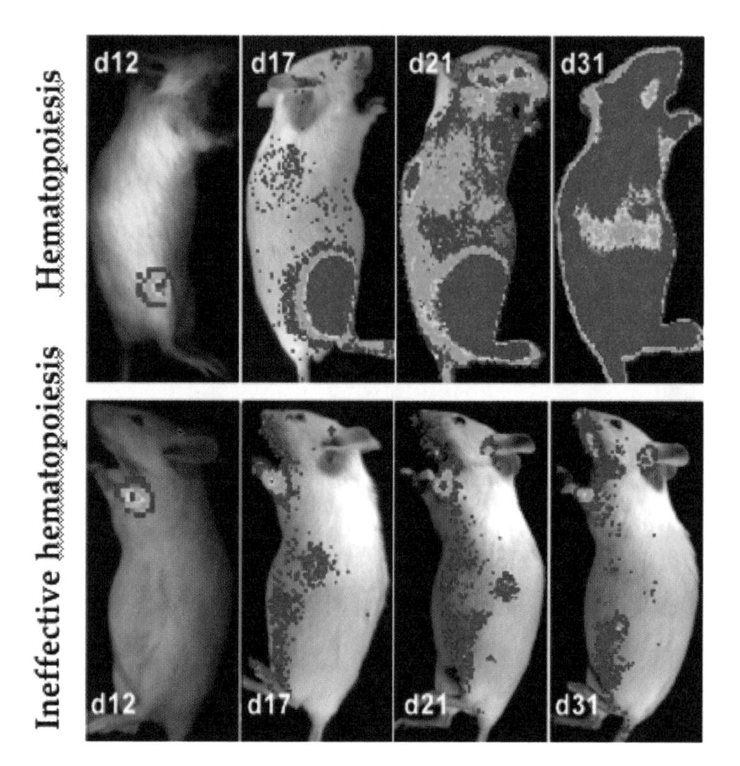

Fig. 2. Temporal analyses of hematopoiesis following single cell transplants. These mice demonstrate the variability in contribution to reconstitution that was observed following the establishment of initial foci of hematopoietic engraftment after transplantation of a single transgenic luc+ hematopoietic stem cells. Pseudo-colored images reflecting optical signal intensity are overlaid upon a grayscale reference image of the animal. (Reprinted from **ref.** ***2***, with permission.)

animal and human subjects, but sensitivity to molecular changes is less than that of other modalities used for these applications. Additionally, approaches for labeling cells are still relatively new *(11)* and their development for detecting labeled cells within the body is ongoing. PET imaging is based on the simultaneous emission of two gamma rays upon annihilation of positrons emitted from radioactive tracers injected into the subject. Resolution in PET is less than that which can be achieved by MRI, but the cross-sectional information and three-dimensional (3D) reconstruction capability offer the potential to be more informative than the typical planar projection data obtained using optical imaging techniques *(12)*. SPECT is based on detection of gamma emission directly from spontaneously decaying radioisotopes that are injected into the subject. It has been used less than PET, perhaps because of its inferior resolution and sensitivity compared with PET. A disadvantage to the use of PET is the necessity for a nearby cyclotron and onsite chemistry to generate the appropriate radioactive labels for labeling cells for injection into research subjects. The decay half-lives for the radioisotopes used in PET are relatively short. This requires that the time from generation of the label to detection in the scanner be minimized to optimize signal strength. Although this is not true for SPECT imaging, both PET and SPECT are somewhat limited by the special handling required for the use of radioisotopes.

Whole-body imaging using bioluminescence and fluorescence is possible for small animal models of human physiology and disease. In general, these technologies are rapid and sensitive and they can be used with molecular markers that are quite versatile. BLI produces planar projection data using pseudo-colored images to represent signal intensity that is localized over grayscale reference images of the subjects (**Figs. 1** and **2**). Optimal detection of bioluminescent or fluorescent signals requires the use of imaging systems that are sensitive to the weak signals that escape the scattering and absorbing environment of the mammalian body. These systems are typically based on charge-coupled device (CCD) detectors and lenses that operate in the visible to near infrared regions of the spectrum. Anatomic resolution is relatively poor in whole-body images, but when necessary, a high magnification lens can be directed at sites in the body where labeled cells have been localized by whole-body imaging to produce high-resolution images that complement the lower resolution images taken noninvasively of the whole animal. This intravital microscopy approach has been published for a number of cell trafficking studies and tremendous insights have been gained in this manner ***(13–16)***. The lack of radioactivity and relatively lower costs make optical modalities more accessible and more available technologies for studies in small animal models.

1.2. Labels

In order to be tracked using any of the imaging modalities, cells of interest must be labeled for detection. A number of studies have used exogenous fluorescent dyes to label cells outside the body prior to transplantation ***(17–19)***. Although these dyes can have a relatively intense signal, a disadvantage of such techniques is that these dyes can be short-lived, and as labeled cells divide, the progeny cells, depending on the dye, may not be labeled and thus the signal is lost over time due to dilution through cell division. Some fluorescent dyes do not produce sufficient signal to be detectable by cameras placed outside the body, necessitating euthanasia of the animal and tissue sampling for analysis. For these reasons, the in vivo imaging techniques, by and large, have required the application of genetic labels. These "labels" are genes that must be introduced into cells and encode proteins that interact with "reporter probes," applied substrates (bioluminescence), or exogenous excitation light sources (fluorescence) to generate a signal that can be localized from outside the body (*see* **Table 1**).

Genetic labels have been developed for all of the modalities used in stem cell trafficking studies. The most commonly used genetic label for PET imaging is the herpes simplex virus thymidine kinase (HSV1-tk). At the time of imaging, [^{18}F]-FHBG or [^{124}I]-FIAU is administered to the animal as a radioactive reporter probe that has specificity for this TK enzyme. The probe is transported into cells and is phosphorylated by the TK protein only in the genetically labeled cells. The phosphorylated probe becomes trapped in the cell and accumulates there, preferentially flagging the genetically labeled cells for detection. The human dopamine D2 receptor (hD2R), the human somatostatin receptor (hSSTR2), and the human transferrin receptor (hTfR) have been used with different imaging modalities and function simply as transmembrane receptors that actively transport their corresponding reporter probes into the genetically labeled cells. In this manner, like HSV1-tk, they also accumulate the probe within the labeled cells to flag them for detection.

Fluorescence and bioluminescence optical imaging modalities require the generation of light by the cells of interest. Genetic labels for BLI are genes that encode luciferases. The "reporter probes" for these proteins are substrates that are oxidized and chemically consumed by the luciferases in reactions that generate light. Fluorescence imaging detects cells that express fluorescent proteins. By analogy to PET and SPECT, the "reporter probe" that must be delivered to these genetically labeled cells is exogenous light of the appropriate wavelength to serve as an excitation source.

In choosing a label for optical imaging, a few key parameters must be considered. Specifically, absorption and substrate biodistribution significantly influence the behavior of the experimental system. In the intact animal, absorption of light by tissue, and in particular absorption by hemoglobin, attenuates the detectable signal generated by cells of interest. Red and infrared light (wavelengths >600 nm) suffers less signal attenuation in the body due to absorption than does light with shorter wavelengths (<600 nm). This is an advantage to firefly luciferase (Fluc; derived from the North American Firefly Photinus pyralis) and click beetle red luciferase (CBRluc; derived from *Pyrophorus plagiophalam*) over Renilla luciferase (Rluc; derived from the sea pansy *Renilla reniformis*)—both Fluc and CBRluc show emission peaks at approx 620 nm at body temperature, whereas the emission peak of Rluc is at 480 nm, in a region where absorption by hemoglobin is relatively high.

The biodistribution of the administered substrate is also an important consideration for interpreting BLI data. To obtain quantitative information from

Table 1
Commonly Used Genetic Labels and Their Corresponding Reporter Probes or Substrates

Modality	Label	Reporter Probe
PET	HSV1-tk	5-[^{124}I]iodo-2′-fluoro-2′deoxy-1-β-D-arabino-furanosyl-uracil (FIAU)
		9-(4-[^{18}F]fluoro-3-hydroxymethylbutyl)guanine (FHBG)
	hD2R	3-(2′-[^{18}F]fluoroethyl)spiperone (FESP)
SPECT	hSSTR2	[^{131}I]DTPA-octreotide
	HSV1-tk	5-[^{131}I]iodo-2'-fluoro-2′deoxy-1-β-D-arabino-furanosyl-uracil (FIAU)
MRI	hTfR	magnetic iron oxides
BLI	Fluc	D-luciferin
	CBRluc	D-luciferin
	Rluc	coelenterazine
FIm	eGFP and derivatives	(excitation source)

BLI data, the substrate must not be limiting in the oxidation reaction that generates light. Therefore, proper imaging technique requires appropriate timing from the administration of substrate to the acquisition of data. The optimal time from administration to acquisition depends upon both the route of administration and the rate of clearance of the substrate in the body. The substrate for Fluc and CBRluc, D-luciferin, is relatively stable in the body and has a relatively long circulation time. In contrast, the substrate for Rluc, coelenterazine, is rapidly cleared from the body and binds to serum proteins ***(20)***. Coelenterazine, therefore, can only be administered intravenously and data acquisition must be complete within a few seconds to a minute after injection. D-Luciferin can be injected intraperitoneally or intravenously ***(21)***. The biodistribution of D-luciferin, after intraperitoneal injection, peaks at 15–20 min and the bioluminescence signal stays relatively stable for another 15–20 min before degrading as a result of substrate clearance. After intravenous injection of D-luciferin the peak of light emission occurs at about 1 min and decreases rapidly ***(21–23)***.

Fluorescence imaging, like detection of any optical signal through mammalian tissues, also requires consideration of light absorption by tissues In this case, the wavelengths of both the excitation source and emitted light must be considered. The most commonly used genetic label for fluorescence imaging is enhanced green fluorescent protein (eGFP). The advantage to using fluorescent labels is the relative ease of detecting their activity in histological samples by fluorescence microscopy, as well as in single-cell suspensions by flow cytometry. These other modalities can provide a satisfying cross-verification to data uncovered using in vivo imaging. The excitation and emission peaks for eGFP occur well below 600 nm and light at these wavelengths is subject to severe signal attenuation by hemoglobin. Imaging of reporters with optical signatures in the blue and green region of the spectrum can be severely constrained by absorption. Whole-body imaging of labeled cells in deep tissues can be difficult, but signals from relatively superficial sites, such as skin and subcutaneous tissues, or from deep sites after removal of overlying tissues, can offer high-resolution images. Several fluorescent proteins with longer emission wavelengths are available (yellow fluorescent protein, dsRed, and others) but these, by and large, have excitation peaks below 600 nm. For these reasons, the use of these labels for deep tissue *in vivo* fluorescence imaging has been shown to require high levels of reporter gene expression from a large number of cells.

1.3. Methods

There are several methods for introducing genetic labels into cells of interest for the purpose of tracking using in vivo imaging techniques. Different cells are variably amenable to the different labeling methods so the appropriate labeling method for any given experimental system will vary based on the cell types involved. This chapter includes detailed protocols for retroviral and lentiviral transduction. Although other methods, such as electroporation and liposome-mediated transfection, have been tried, viral transduction and transgenesis have proven to be to most effective methods for labeling bone marrow cells.

Labeling cells using classical transgenesis involves the generation of a plasmid encoding the genetic label of interest driven by an appropriate promoter. The plasmid is stably transfected (using conventional methods such as electroporation or liposome-mediated or calcium phosphate mediated transfection) into an embryonic stem cell line and those cells are microinjected into murine blastocysts. The blastocysts are transferred into pseudo-pregnant females and chimeric pups are selected and bred until the transgene is found in all progeny tissues. This process can be time-consuming and labor intensive, and usually requires the technical expertise of an available transgenic core facility. Genetic labels applied in this way exhibit a relatively predictable expression pattern but can be subject to the vagaries of random chromosomal integration. They may integrate into a site that is subject to transcriptional regulation that is not dictated by the originally designed plasmid. This is a so-called "founder effect" or "contextual influence" and is best screened for by examining the behavior of the transgene in multiple sibling chimeric lines.

Viral transduction is accomplished by replication-incompetent viral vectors that insert the genetic label directly into the genome of the target cell ***(24)***. The genetic label is first cloned into a plasmid that contains the viral packaging sequence (ψ) and promoter sequences (long terminal repeat [LTR] sequences). The plasmid is then transfected into a packaging cell line. The packaging line is a cell line that has been stably transfected with genes for viral envelope and packaging proteins. When these proteins are present in the same cell as DNA flagged with the ψ packaging sequence, viral particles are generated that contain the genetic label. Because the genes encoding envelope and packaging proteins are not present in the segment of DNA that is flagged by the viral packaging sequence, they are not included in the resulting viral particles. Those particles are, thus, replication-incompetent. Once they have infected a target cell, they insert the appropriate DNA into the target genome and then, because no packaging or envelope proteins are present in the target cell, no further virus is generated.

An alternative to using a stably transfected "viral packaging cell line" is to use a second plasmid that contains only the genes encoding envelope and packaging proteins. This second plasmid can be co-transfected along with the plasmid containing the genetic label into an unmodified highly transfectable cell line of choice (e.g., 293T, as the lentiviral transduction protocol below uses.) In the cells that receive both plasmids, the presence of ψ-flagged DNA along with envelope and packaging proteins leads to the generation of viral particles. As with the stably transfected packaging line, the fact that the genes encoding envelope and packaging proteins are not present on the ψ-flagged DNA means that those genes will not be included in the viral particles and will not be present in the target cell. Again, no virus will be generated in the target cell, but the genetic label will be inserted into its genome.

Lentiviral and retroviral transduction protocols are available and an example of each is included in this chapter. The retroviral transduction protocol is optimized for mouse bone marrow and stem cells and the lentiviral transduction protocol is specific for sorted human hematopoietic stem cells (HSCs). Both human and mouse cells can be transplanted into mice for optical imaging of cell trafficking patterns, but the human cells must be transplanted into immunocompromised mice (e.g., severe combined immunodeficient [SCID]-Hu.) Although lentiviral transduction can be used to label murine bone marrow cells, retroviral transduction systems have been used more extensively with mouse cells and the reagents are readily available. Human HSCs are less amenable to retroviral transduction. Currently, the best success with transduction of human HSCs has been achieved using lentiviral systems.

The disadvantage to using retroviral transduction is that retroviral particles are only able to insert their DNA into the genomes of cells that are actively dividing. Protocols for retroviral labeling of bone marrow cells are thus carefully optimized for activation and stimulation of HSCs while preventing lineage differentiation. Lentiviral transduction systems do not require the target cells to be in cycle so are more appropriate for labeling cells that may not be actively dividing. However, if the target cells are HSCs, the same issues of maintaining the cells in culture while preventing lineage differentiation still arise. Genetic labels applied through viral transduction (either retroviral or lentiviral) have the advantage of being easier to apply than transgenic labels. The disadvantage is the unpredictable impact of cell culture and viral transduction on the biological behavior of the target cells.

2. Materials

2.1. Retroviral Transduction of Murine Bone Marrow and Stem Cells

1. 5-Fluorouracil (5-FU; for human injection), American Pharmaceutical Partners, Inc. NDC 63323-117-10 (50 mg/mL, 10 mL). Stock concentration 25 mg/mL in phosphate-buffered saline (PBS), aliquot, and store at –20°C.
2. Retroviral packaging cell line.
3. Retroviral expression plasmid.
4. Lipofectamine 2000 transfection reagent (Invitrogen, cat. no. 11668-027) (0.75 mL) (*see* **Note 1**).
5. Opti-MEM I reduced serum medium (Invitrogen, cat. no. 31985-070) (500 mL).
6. RBC lysing buffer (Sigma, cat. no. R7757) (100-mL bottle).
7. rmIL-3 (Peprotech, cat. no. 213-13) (10 μg). Stock concentration 100 μg/mL in sterile water, aliquot and store at –20°C.
8. rmIL-6 (Peprotech, cat. no. 216-16) (10 μg). Stock concentration 100 μg/mL in sterile water, aliquot and store at –20°C.
9. rmSCF (Peptrotech, cat. no. 250-03) (10 μg). Stock concentration 100 μg/mL in sterile water, aliquot and store at –20°C.
10. Fetal calf serum, heat-inactivated to 56°C for 45 min (Invitrogen, cat no. 16000-044) (500 mL).
11. Dulbecco's modified Eagle's medium (DMEM; Invitrogen, cat. no. 10569-010) (500 mL), or StemSpan SFEM (Stem Cell Technologies, cat. no. 09650) (500 mL).
12. Hexadimethrine bromide (polybrene; Aldrich, cat. no. 107689-100G). Stock concentration 5 mg/mL in PBS, store at 4°C.
13. HEPES Buffer Solution (1 *M*; Invitrogen, cat. no. 15630-080) (100 mL).

2.2 Lentiviral Transduction of Human HSCs and Progenitor Cells

1. Highly transfectable cell line to be used as packaging cells. This protocol describes conditions optimized for 293T cells.

2. Growth media for packaging cells. For 293T: DMEM with 10% heat inactivated fetal bovine serum, 100 U/mL penicillin and 100 U/mL streptomycin, 2 m*M* L-glutamine, 50 μg/mL 2-mercaptoethanol.
3. Calcium Phosphate Transfection Kit (Invitrogen, cat. no. K2780-01) (*see* **Note 2**).
4. Lentiviral expression plasmid (e.g., pWPTS, pWPXL, pWPI, pLVTH).
5. Lentiviral packaging plasmid (e.g., pCMV-dR8.91, pCMV-dR8.74, psPAX2).
6. Lentiviral envelope plasmid (e.g., pMD2G).
7. Hexadimethrine Bomide (polybrene; Aldrich, cat. no. 107689-100G). Stock concentration 5 mg/mL in PBS, store at 4°C.
8. Myelocult H5100 media (Stem Cell Technologies, cat. no. 05150) (500 mL).
9. rh interleukin (IL)-6 (R&D Systems, cat. no. 206-IL-050) (50 μg).
10. rh TPO (R&D Systems, cat. no. 288-TP-025) (25 μg).
11. rh Flt-3L (R&D Systems, cat no. 308-FKN-025) (25 μg).
12. rh SCF (R&D Systems, cat. no. 255-SC-050) (50 μg).

2.3. Bone Marrow Transplantation Into Mice

1. Irradiation source for use with mice.
2. Mouse irradiation chamber (Shadel, Inc., cat. no. 62419).
3. Labeled bone marrow cells in PBS.
4. 28-gage, one-half-inch, 1 cc/U-100 insulin syringes (Becton-Dickenson, cat no. 329424; one box).
5. Heating lamp/restraining device.
6. Trimethoprim/Sulfamethoxazole suspension (40 mg TMP and 200 mg Sulfa/5 cc), (Hi-Tech Pharmacal, cat. no. NDC 50383-823-16) (200 mL).

2.4. Bioluminescence Imaging (see **Note 3**)

1. Low-light imaging system for BLI.
2. Anesthetic agent.
3. Substrate:
   a. Luciferin (Biosynth, Int., cat. no. L-8220; 1 g). Stock concentration 30 mg/mL in PBS, aliquot and store at –20°C.
   b. Coelenterazine (Nanolight, cat. no. NFCTZFB) (5 mg). Stock concentration 10 mg/mL in ethanol, aliquot and store at –80°C. For working stock, dilute frozen stock 10 μL into 1.5 mL of PBS. Do not store.

3. Methods

3.1. Retroviral Transduction of Murine Bone Marrow and Stem Cells

Day 1:

1. Inject 250 mg/kg 5FU intraperitoneally or intravenously into each donor mouse (*see* **Note 4**).

Day 3:

2. Split packaging cell line for transfection per protocol. For Lipofectamine 2000, seed 1–1.5 × 10^6 cells per 6-cm dish. Incubate at 37°C/5% CO_2 overnight.

DAY 4:

3. Transfect packaging cell line per protocol. For Lipofectamine 2000, prepare DNA-Lipofectamine 2000 complexes, one tube for each sample:
   a. Dilute 8 μg DNA in 0.5 mL of Opti-MEM I reduced serum medium. Mix gently.
   b. Mix Lipofectamine 2000 gently before use, then dilute 20 μL in 0.5 mL Opti-MEM. Mix gently and incubate for 5 min at room temperature. The next step must be done within 30 min.
   c. After 5–30 min of incubation, combine the diluted DNA with the diluted Lipofectamine 2000 reagent. The total volume should now be 1 mL. Mix gently and incubate for 20 min to 6 h at RT.
4. Add the prepared 1 mL of DNA-Lipofectamine 2000 complexes to each 6-cm plate containing cells and 5 mL of medium. Mix gently by rocking back and forth.
5. Incubate the cells at 37°C/5% CO_2 for 48 h before harvesting viral supernatant.

DAY 5:

6. Euthanize donor mice according to approved animal protocols.
7. Harvest bilateral femurs and tibias from each mouse. Scrape bones with a razor blade to remove any attached muscle, cartilage, and connective tissue. Using razor blade or scissors, remove the ends of the bones (*see* **Note 5**).
8. Using a 3-cc syringe and a one-half-inch, 25-gage needle, insert the needle into each cut end of each bone and squirt the marrow from the center using sterile PBS. Squirt into a sterile dish and refill the syringe with fresh PBS if necessary. Squirt the marrow until the bones turn white.
9. Make a single-cell suspension of marrow cells by aspirating the cells in the sterile dish once through the one-half-inch, 25-gage needle and then pushing them back out through the needle and into a sterile conical tube.
10. Spin at 1200 rpm (approx 310*g*) for 5 min at 4°C.
11. Aspirate supernatant. Resuspend in 3–5 mL RBC lysis buffer (depending on the size and "redness" of the cell pellet).
12. Incubate at room temperature for 10 min.
13. Spin at 1200 rpm (approx 310*g*) for 5 min at 4°C.
14. Resuspend in activation medium (DMEM or StemSpan SFEM + 10% HI FCS + 10 ng/mL IL-3 +50 ng/mL IL-6 + 100ng/mL SCF), 1.5×10^6 cells/mL (*see* **Note 6**).
15. Plate at 1.5×10^6 cells/well in a 24-well plate, 1 mL per well. Incubate at 37°C/ 5% CO_2 overnight.

DAY 6:

16. Remove the supernatant from the transfected packaging cell line. Replace with fresh media and return the cells to the incubator.
17. To the supernatant, add polybrene to 5 μg/mL and HEPES to 10 m*M*. Mix thoroughly.
18. Filter supernatant through a 0.45-μm filter.

19. Centrifuge the bone marrow cells at 2100 rpm (approx 950*g*) for 5 min. Aspirate media.
20. Add 1 mL of filtered viral supernatant to each well of bone marrow cells.
21. Centrifuge the bone marrow cells at 2100 rpm (approx 950*g*) for 75 min at room temperature. Aspirate media.
22. Add 1 mL activation media to each well of bone marrow cells. Return the cells to the incubator.

DAYS 7 AND 8:

23. Repeat d 6 protocol. On day 8, discard transfected packaging cells after removing the viral supernatant.

DAY 9:

24. If there is a fluorescent protein, assess transduction efficiency using flow cytometry.
25. If transduction is adequate, decide on a cell dose and proceed to transplantation as described next (*see* **Note 7**). For tail vein injection, plan for an injection volume of 200–500 µL per mouse.

3.2. Lentiviral Transduction of Human HSCs and Progenitor Cells

DAY 1:

1. Seed 2–3 × 10^6 packaging cells per 10-cm dish in 10 mL of growth media. Incubate at 37°C/5%CO_2 for 24 h. The cells should be 80% confluent at the time of transfection.

DAY 2:

2. Remove the media from the packaging cells and wash the cells with serum free media. Leave the cells under 5 mL of serum-free media.
3. For each 10-cm plate of cells to be transfected, dispense 0.5 mL of 2X HBS buffer (found in calcium phosphate transfection kit) in a 15-mL tube.
4. In a separate 15-mL tube, dispense 100 µL of 2 *M* $CaCl_2$ (also found in the kit.) Add 11 µg lentiviral expression plasmid, 3 µg lentiviral packaging plasmid, and 6 µg of lentiviral envelope plasmid. Add deionized H_2O to a final volume of 500 µL.
5. Slowly add the DNA/calcium mixture drop–wise to the 2X HBS solution while gently vortexing the 15-mL tube. The mixture must not spill.
6. Gently vortex for 10–15 min to allow the formation of calcium phosphate/DNA precipitates.
7. Remove the packaging cells from incubator at the last moment. Add the mixture one drop at a time to the supernatant. Gently rock the culture dish to mix and return the cells to the incubator overnight.

DAY 3:

8. Check the cells under a microscope. Small black particles should be visible.
9. Remove the transfection media. Replace with 10 mL fresh growth media.

DAY 5:

10. Remove the supernatant and filter through 0.45-μm pores.
11. Centrifuge the supernatant at 22,000*g* for 2 h to concentrate the virus (*see* **Note 8**).
12. Resuspend the viral pellet in 250 μL/10-cm plate of packaging cells. Use myelocult media.
13. Test the concentrated virus for optimal dilution by adding it to the supernatant of 293T cells at dilutions of 1:1, 1:10, and 1:100. Plan to use the dilution that gives maximal transduction with minimal cell toxicity.
14. Sort the desired number of HSCs and progenitor cells into Eppendorf tubes, each containing 200 μL of PBS with 2% calf serum. Spin at 1900 rpm in a microfuge for 5 min.
15. Remove the supernatant, then resuspend in myelocult media supplemented to a final concentration of 50 ng/mL SCF, 10 ng/mL IL-6, 10 ng/mL TPO, and 50 ng/mL Flt3 ligand. Keep in mind the volume of virus that will be added. Use a volume appropriate to cover the cells lying in a single cell layer on the bottom of a tissue culture dish.
16. Add the concentrated lentiviral supernatant, to the dilution determined in **step 13** above. Remove the cells to a tissue culture dish with adequate surface area to allow the cells to lie in single cell thickness at the bottom. Incubate at 37°C/5% CO_2 for 24 h.

DAY 6:

17. If there is a fluorescent protein, assess transduction efficiency using flow cytometry.
18. If transduction is adequate, proceed to transplantation as described later. Cell doses of 10,000–50,000 labeled cells into SCID-Hu mice have been used.

3.3. Bone Marrow Transplantation Into Mice

1. Transfer the recipient mice to the mouse irradiation chamber for lethal irradiation.
2. Deliver 800–900 cGy total body irradiation (*see* **Note 9**).
3. Warm the recipient mice under a heating lamp to encourage peripheral vasodilation.
4. One mouse at a time, transfer the mice into a restraining device for tail vein injection (*see* **Note 10**) and inject a rescuing dose of bone marrow cells.
5. After injection, mice should be maintained in autoclaved cages and bedding with autoclaved water supplemented with approximately two capfuls of Trimethoprim/Sulfamethoxazole per full water bottle (*see* **Note 11**). These living conditions should continue for the first month.

3.4. Bioluminescence Imaging Considerations

1. Using an approved method of anesthesia, anesthetize the mice for imaging. Expect to need them to lie still for up to 15 min for multiple images using 1- to 5-min data acquisition times.

2. Initialize the camera. The user interface of Living Image has a command to initialize which resets all parameters.
3. Before imaging, administer the appropriate luciferase substrate by the chosen route. For luciferin, the dose for an adult mouse is usually 150 mg/kg body weight given intraperitoneally. For coelenterazine, this is typically 6.7 μg injected intravenously (*see* **Note 12**).
4. Approximately 15 min after an intraperitoneal injection of luciferin or immediately after intravenous injection of any substrate, take the initial image (In Living Image, click on the "ACQUIRE" button), keeping in mind the following considerations:
   a. Stage height should be adjusted such that the field of view fits the number of mice in the image. It should be at the highest setting that allows visualization of all mice to be imaged. In Living Image, the "ACQUIRE CONTINUOUS PHOTOS" button is helpful for adjusting this. It takes photographic images only.
   b. Shorter exposure times (e.g., 1 s) are appropriate if the signal is expected to be strong. Longer times (e.g., 5–10 min) are required if the signal is predicted to be weak. If the signal strength is unknown, a 1-min initial exposure time is a reasonable place to start. These parameters are set in the user interface.
   c. Binning pixels can be done with extremely weak signals to improve the apparent contrast. The total number of photons collected does not change when binning the pixels, so it is not an increase in sensitivity, but binning does result in a loss of resolution as there is a smaller effective number of elements on the chip. Smaller binning numbers correspond to higher image resolution and lower contrast. Higher binning numbers correspond to better apparent contrast and lower image resolution. Medium binning is a good place to start.
   d. "Saturated" images are nonquantitative, as the pixels that are saturated cannot collect more signal. The previously described parameters should be adjusted to collect an image that reflects a signal strength that is adequate for anatomic localization but not strong enough to saturate the camera.
5. In order to verify and localize a given signal, mice should be imaged in multiple positions. Conventional views include ventral, dorsal and two lateral images.

4. Notes

1. Any transfection protocol can be used. This protocol will use the Lipofectamine 2000 conditions.
2. These 5FU doses are optimized for B6 mice. Doses of 150 mg/kg have also been used with 48 h pretransduction activation times in vitro.
3. Try to remove as little bone as possible, as most of the HSCs are found near the ends of the bones.
4. Some have found that StemSpan SFEM-based media supplemented with calf serum and cytokines results in superior viability and activation of the bone marrow cells.
5. The cell dose required to rescue will vary depending on the quality of the 5FU enrichment as well as the quality of the response to the activation media. Plan on

being able to rescue one or two mice per well, depending on the health of the cells in culture as assessed by light microscopy.

6. Both calcium phosphate and liposome-mediated transfection methods have been used. This protocol describes a modified calcium phosphate-based method.
7. Virus can be harvested again from the transfected packaging cells in 72 h. After collection, virus can be concentrated as previously listed. Discard the supernatant and store the viral pellet in the Eppendorf tube at –80°C.
8. Lethal doses of total body irradiation vary by background strain. Balb/C and FVBN mice typically use 800 cGy, whereas C57B6 mice require 900 cGy. For C57B6, the dose can also be fractionated into an 800-cGy dose followed by a 400-cGy dose approx 3 h later.
9. Rescuing bone marrow cells can also be introduced by retro-orbital injection. In that case, the heating lamp and restraining device are not necessary but anesthesia may be required, and the injection volume should be 100–200 μL per mouse.
10. A mixture of Neomycin (1.2 g/L) and Polymixin-B (1×10^6 U/L) in the drinking water can also be used.
11. Several options for imaging systems are commercially available. This protocol describes the use of the Xenogen IVIS system and associated Living Image software.
12. Biodistribution and ultimately consumption of the substrate throughout the mouse will result in the bioluminescence signal strength changing over time. For luciferin given intraperitoneally, the signal increases slowly and appears to plateau at approx 15–20 min after the injection. It remains relatively stable for an additional 15–20 min and then declines. The precise timing of the plateau in signal should be measured in each experimental system by taking several consecutive images and noting the time from injection to signal stability. Luciferin can also be given intravenously, and this is helpful for reaching anatomic sites such as the brain, but clearance of the substrate will occur much more rapidly than when substrate is delivered intraperitoneally.

References

1. Hardy, J. F. K., DeBoer, M., Chu, P., Gibbs, K., and Contag, C. H. (2004) Extracellular replication of Listeria monocytogenes in the murine gall bladder. *Science* **303,** 851–853.
2. Cao, Y. W. A., Beilhack, A., Dusich, J., et al. (2004) Shifting foci of hematopoiesis during reconstitution from single stem cells. *PNAS* **101,** 221–226.
3. Blasberg R. G. and Gelovani, J. (2002) Molecular-genetic imaging: a nuclear medicine-based perspective. *Mol. Imaging* **1,** 280–300.
4. Bennink R. J., Hamann. J., de Bruin, K., ten Kate, F. J., van Deventer, S. J., and te Velde, A. A. (2005) Dedicated pinhole SPECT of intestinal neutrophil recruitment in a mouse model of dextran sulfate sodium-induced colitis. *J. Nucl. Med.* **46,** 526–531.
5. Bennink, R. J., van Montfrans, C., de Jonge, W. J., de Bruin, K., van Deventer, S. J., and te Velde, A. A. (2004) Imaging of intestinal lymphocyte homing by

means of pinhole SPECT in a TNBS colitis mouse model. *Nucl. Med. Biol.* **31,** 93–101.

6. Hildebrandt, I. J. and Gambhir, S. S. (2004) Molecular imaging applications for immunology. *Clin. Immunol.* **111,** 210–224.
7. Jendelova, P., Herynek, V., Urdzikova, L., et al. (2004) Magnetic resonance tracking of transplanted bone marrow and embryonic stem cells labeled by iron oxide nanoparticles in rat brain and spinal cord. *J. Neurosci. Res.* **76,** 232–243.
8. Modo, M., Cash, D., Mellodew, K., et al. (2002) Tracking transplanted stem cell migration using bifunctional, contrast agent-enhanced, magnetic resonance imaging. *Neuroimage* **17,** 803–811.
9. Stroh, A., Faber, C., Neuberger, T., et al. (2005) In vivo detection limits of magnetically labeled embryonic stem cells in the rat brain using high-field (17.6 T) magnetic resonance imaging. *Neuroimage* **24,** 635–645.
10. Wang, X., Rosol, M., Ge, S., et al. (2003) Dynamic tracking of human hematopoietic stem cell engraftment using in vivo bioluminescence imaging. *Blood* **102,** 3478–3482.
11. Cunningham, C. H., Arai, T., Yang, P. C., McConnell, M. V., Pauly, J. M., and Conolly, S. M. (2005) Positive contrast magnetic resonance imaging of cells labeled with magnetic nanoparticles. *Magn. Reson. Med.* **53,** 999–1005.
12. Kim, Y. J., Dubey, P., Ray, P., Gambhir, S. S., and Witte, O. N. (2004) Multimodality imaging of lymphocytic migration using lentiviral-based transduction of a tri-fusion reporter gene. *Mol. Imaging Biol.* **6,** 331–340.
13. Delon, J., Stoll, S., and Germain, R. N. (2002) Imaging of T-cell interactions with antigen presenting cells in culture and in intact lymphoid tissue. *Immunol. Rev.* **189,** 51–363.
14. Sipkins, D. A., Wei, X., Wu, J. W., et al. (2005) In vivo imaging of specialized bone marrow endothelial microdomains for tumour engraftment. *Nature* **435,** 969–973.
15. Sorg, B. S., Moeller, B. J., Donovan, O., Cao, Y., and Dewhirst, M. W. (2005) Hyperspectral imaging of hemoglobin saturation in tumor microvasculature and tumor hypoxia development. *J. Biomed. Opt.* **10,** 44,004.
16. Stoll, S., Delon, J., Brotz, T. M., and Germain, R. N. (2002) Dynamic imaging of T cell-dendritic cell interactions in lymph nodes. *Science* **296,** 1873–1876.
17. Cui, J., Wahl, R. L., Shen, T., et al. (1999) Bone marrow cell trafficking following intravenous administration. *Br. J. Haematol.* **107,** 895–902.
18. Krause, D. S., Theise, N. D., Collector, M. I., et al. (2001) Multi-organ, multi-lineage engraftment by a single bone marrow-derived stem cell. *Cell* **105,** 369–377.
19. Szilvassy, S. J., Meyerrose, T. E., Ragland, P. L., and Grimes, B. (2001) Differential homing and engraftment properties of hematopoietic progenitor cells from murine bone marrow, mobilized peripheral blood, and fetal liver. *Blood* **98,** 2108–2115.
20. Zhao, H., Doyle, T. C., Wong, R. J., et al. (2004) Characterization of coelenterazine analogs for measurements of Renilla luciferase activity in live cells and living animals. *Mol. Imaging* **3,** 43–54.

21. Zhao, H., Doyle, T. C., Coquoz, O., Kalish, F., Rice, B. W., and Contag, C. H. (2005) Emission spectra of bioluminescent reporters and interaction with mammalian tissue determine the sensitivity of detection in vivo. *J. Biomed. Opt.* **10,** 41,210.
22. Bhaumik, S. and Gambhir, S. S. (2002) Optical imaging of Renilla luciferase reporter gene expression in living mice. *PNAS* **99,** 377–382.
23. Bhaumik, S., Lewis, X. Z., and Gambhir, S. S. (2004) Optical imaging of Renilla luciferase, synthetic Renilla luciferase, and firefly luciferase reporter gene expression in living mice. *J. Biomed. Opt.* **9,** 578–586.
24. Miller, A. (1997) Development and applications of retroviral vectors, in *Retroviruses* (Coffin, J. M., Hughes, S. H., and Varmus, H. E., ed.), CSHL, Woodbury, NY, pp. 437–473.

3

An Overview of HLA Typing for Hematopoietic Stem Cell Transplantation

Ann-Margaret Little

Summary

Selection of a related or unrelated haematopoietic stem cell donor for a patient requires accurate matching of human leukocyte antigen (HLA) genes in order to maximise the beneficial effects of the transplant. There are a number of different approaches that can be made in order to achieve HLA type depending on the number of samples being processed, the level of resolution to be achieved, and the cost of providing the various tests. Each method has its advantages and disadvantages and in most laboratories, a combination of methods may be used.

Key Words: HLA typing; PCR-SSP; PCR-SSO; PCR sequencing.

1. Introduction

Polymorphic human leukocyte antigen (HLA) genes are encoded within the human major histocompatibility complex on chromosome 6 (*see* **Table 1**). These genes encode components of HLA molecules involved in antigen presentation to the immune system. Determination of HLA polymorphism (referred to as "HLA typing" or "tissue-typing") is performed in clinical laboratories in order to aid matching of potential organ and stem cell donors for patients and also as an aid in diagnosis of various diseases that are associated with a particular HLA genotype *(1)*. In addition, polymorphisms within HLA genes and other genes linked to HLA genes within the major histocompatibility complex are now known to play a role in the responsiveness of individuals to vaccines and other pharmaceutical drugs *(2)*. HLA typing is also used as a tool to aid genetic and anthropological studies of human populations *(3)*.

HLA typing methods have evolved significantly since the discovery of the first HLA antigens from studies with sera from multiparous women *(4)*. Originally serological and cellular methods were applied to determine HLA type.

From: *Methods in Molecular Medicine, vol. 134: Bone Marrow and Stem Cell Transplantation*
Edited by: M. Beksac © Humana Press Inc., Totowa, NJ

Table 1
Human Leukocyte Antigen (HLA) Genes and Number of Alleles, February 2007

HLA locus	No. of alleles	No. of expressed proteins
HLA-A	506	405
HLA-B	851	729
HLA-C	276	219
HLA-DRA	3	2
HLA-DRB1	476	404
HLA-DRB3 (DR52)	44	36
HLA-DRB4 (DR53)	13	7
HLA-DRB5 (DR51)	18	15
HLA-DQA1	34	25
HLA-DQB1	81	59
HLA-DPA1	23	14
HLA-DPB1	126	113

See **ref.** ***23***.

However, since the application of DNA-based methods, particularly after the introduction of the PCR, the sophistication of the methods available for HLA typing has increased significantly. This chapter will focus on the currently available DNA-based methods, their applicability for different tests and their advantages and disadvantages.

2. HLA Typing Methods

The many HLA alleles found are the product of multiple single-nucleotide polymorphisms (SNPs) distributed singly or in groups "motifs" via recombinatorial mechanisms generating what is often referred to as a "patchwork" pattern of polymorphism. This extensive polymorphism of HLA makes this genetic system one of the most complex to analyse. The three methods most widely applied for HLA typing use PCR as a starting point.

1. PCR using sequence-specific oligonucleotides (PCR-SSO).
2. PCR using sequence-specific primers (PCR-SSP).
3. PCR sequencing.

2.1. PCR-SSO

PCR-SSO involves a PCR reaction producing an amplicon from the HLA locus to be tested, e.g., HLA-A or from a group of related HLA alleles from the same locus, e.g., HLA-DRB1*04. The PCR product is then hybridised against a panel of oligonucleotide probes that have sequences complementary to stretches of polymorphisms within the target HLA alleles (*see* **Fig. 1**). The process of hybridization has changed significantly since the first descriptions of this method. Originally, denatured PCR products (produce single-stranded DNA) were immobilized to multiple nitrocellulose or nylon membranes, the number of membranes equating with the number of oligonucleotide probes to be used. After hybridization of each membrane with a different oligonucleotide probe (which may involve different hybridization temperatures for different probes), bound probes were visualized using detection systems such as radio-isotopes or later chemiluminescence. Such systems were (and still are) widely used for HLA typing multiple individuals in one "batch" *(5)*. The number of PCR products that could be immobilized on a membrane depends in part on the automation available for accurate "dotting" of PCR products and also the available software used for the final discrimination and analysis of the positive and negative reactions. Therefore this methodology is very amenable to high-throughput HLA typing as is required for busy hematopoietic stem cell donor registries. The disadvantages of PCR-SSO include the labor-intensiveness of the method, which can take several days to perform from processing PCR product to final result. The requirement for multiple hybridization temperatures for different oligonucleotide probes also meant use of several water baths, thus taking up valuable laboratory space. The introduction of tetramethylammonium chloride (TMAC) in hybridization buffers allowed the hybridization temperatures to become more uniform; however, the use of TMAC was not welcome because of its toxic qualities. Although this method of PCR-SSO is still in use, it has mostly been superseded by the introduction of "reverse" PCR-SSO methods.

The reverse PCR-SSO methods involve immobilization of the oligonucleotide probes on a solid-phase support, and subsequent hybridization of the solid-phase immobilized oligonucleotide probes with liquid-phase PCR product. Solid-phase supports most widely in use include:

1. Nitrocellulose membrane strips (Reli™ SSO Dynal Biotech, The Wirral, UK and INNOLiPA line probe assay, Innogenetics, Gent, Belgium).
2. Microtest plates (enzyme-linked probe hybridization assay [ELPHA], Biotest, Germany).
3. Oligonucleotide-coated polystyrene microspheres (Luminex® xMAP technology, Luminex Corp, Austin, TX; LIFEMATCH™, Tepnel Lifecodes, Stamford, CT; and LABType® SSO, One Lambda Inc., Los Angeles, CA).

The establishment of the reverse PCR-SSO allowed the introduction of this method for protocols in addition to large-throughput HLA typing. As hybridization conditions are standardized for all oligonucleotide probes immobilized, the technical demands on the technology are reduced and there is greater capacity for automation.

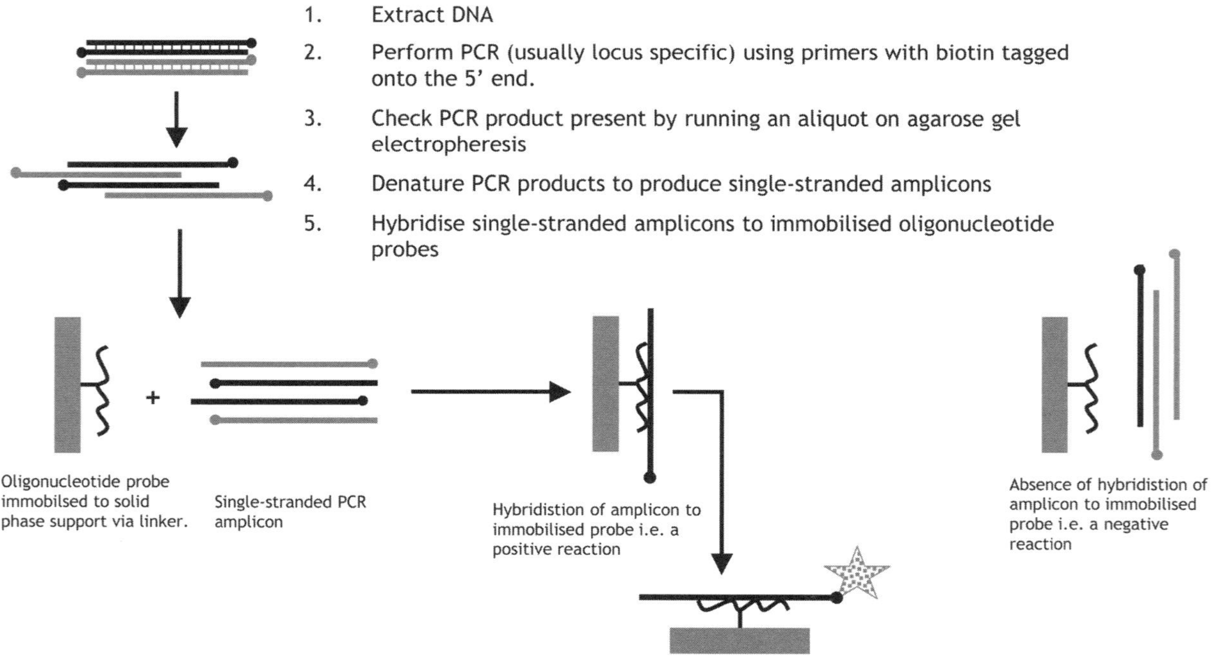

1. Extract DNA
2. Perform PCR (usually locus specific) using primers with biotin tagged onto the 5’ end.
3. Check PCR product present by running an aliquot on agarose gel electropheresis
4. Denature PCR products to produce single-stranded amplicons
5. Hybridise single-stranded amplicons to immobilised oligonucleotide probes
6. Addition of streptavidin conjugated enzyme which emits colour after addition of substrate allows identification of positive reactions
7. Pattern of positive and negative reactions are interpreted to give HLA type

HLA locus	No. probes
HLA-A	61
HLA-B	96
HLA-DRB1	70

Example of number of probes used to obtain medium level resolution in a commercially available kit.

Further solid-phase supports in development include microarray technology. The microarray technology offers the ability to immobilize a greater number of either PCR products (direct SSO) or oligonucleotides (reverse SSO) on microarray slides (~10,000 per slide) compared with the conventional methods described previously ***(6)***. Further adaptations of the PCR-SSO include combining hybridisation steps with ligation steps in the "Universal Array" technology. This allows the use of an array system with immobilised probes that are not specific for the target polymorphisms, i.e., are universal for all polymorphisms. The target sequence to be analyzed is produced by PCR and incubated with three probes: two allele-specific oligonucleotides (one specific for wild-type sequence, the other specific for mutant sequence), and a third probe that targets a sequence common to both mutant and wild-type alleles. The allele specific probes have their 3′-end complementary to the target polymorphism and the common sequence probe has its 5′-end (with free 5′ phosphate) adjacent to the 3′-end of the allele specific probes. The common allele also possesses a "reporter" sequence at its 3′-end which is complementary to a specific

---

Fig. 1. *(opposite page)* PCR using sequence-specific oligonucleotides (PCR-SSO) simplified. After DNA extraction (1), locus-specific PCR product is produced (2), e.g., human leukocyte antigen (HLA)-A. The black and grey double-stranded PCR products indicate the presence of two alleles in the amplicon, e.g., HLA-A*01 and A*02. The circles at the ends of each strand represent the incorporation of biotin on the PCR primers. The presence of correct sized PCR product without contamination is determined by separating an aliquot of the amplicon by agarose gel electrophoresis (3). The PCR amplicon is denatured (chemically or heat) to produce single-stranded DNA for both sense and antisense strands (4). The single-stranded amplicons are hybridized to the immobilized oligonucleotide probe, which is attached to a solid support. e.g.. membrane, microplate well, bead. Amplicons that do not hybridize are washed away (5). The numbers in the table gives the number of probe reactions that require interpretation (i.e., as positive or negative) for the LAB-Type® SSO HLA typing tests to give a medium resolution type (Lots 006, 008 and 009 for HLA-A, B and DRB1, respectively).

The presence of bound amplicon to oligonucleotide probe is measured using an enzymatic colour reaction, e.g., addition of streptavidin conjugated to an enzyme, e.g., horseradish peroxidase or alkaline phosphatase which will catalyze the formation of a colored complex in the presence of appropriate substrate or alternatively the addition of fluorescently labeled streptavidin allows identification of bound PCR product (6). Interpretation of the results is the most complex aspect of the PCR-SSO test and requires the use of specific software if all known HLA alleles are included in the analysis.

immobilized universal probes. In the presence of DNA ligase, the common probe will ligate to bound specific allele probe hybridized to the target sequence and will hybridize to the immobilized universal probe *(7)*. This can be detected by the attachment of different fluorescent dyes to the two allele-specific probes.

One of the limitation of PCR-SSO methods has been the inability to link *cis* and *trans* polymorphisms. However, with the design of longer probes that discriminate multiple polymorphisms, this problem can be reduced (*see* **Figs. 2A** and **3**).

Elongation multiplexed analysis of polymorphisms (eMAP) can also be applied to link *cis* polymorphisms. The principle of this assay involves the hybridization of an oligonucleotide primer to a target polymorphism, extension of the nucleotide sequence from the oligonucleotide primer, followed by hybridization of a second oligonucleotide (specific for the linked polymorphism) to the extended sequence. Such an approach has been applied to HLA typing *(8)*.

2.2. PCR-SSP

The other most widely used DNA methodology for HLA typing is PCR-SSP. PCR-SSP involves the use of multiple PCR reactions targeting the presence and absence of polymorphisms within the target HLA gene *(9,10)*. Each PCR primer pair is designed to be either complementary to the target polymorphism or is designed to be a mismatch through substitution of the complementary nucleotide at the 3′-end of either or both primers with a noncomplementary nucleotide. This approach is also called the amplification refractory mutation system (ARMS) *(11)*. As the Taq polymerase enzyme used in the PCR reaction does not have a 3′ to 5′ exonuclease proofreading ability, mismatched priming does not result in production of a PCR amplicon. In order for this approach to be successful and for PCR-SSP to be used routinely for HLA typing, it is necessary to optimise the annealing and other PCR conditions including buffer components, for each primer pair utilized. The amplicons produced, which are indicative of the presence of the target polymorphism within the sample DNA being tested, are visualized after agarose gel electrohoresis and staining, usually with ethidium bromide. Most PCR-SSP protocols include, within each amplification mix, primers for amplification of a segment from a housekeeping gene. The presence of this additional amplicon serves as an internal control, as it should be amplified from all human samples tested **(Fig. 4A,B)**.

PCR-SSP is an extremely flexible method that can be performed in laboratories with minimum molecular biology equipment (thermal cycler, agarose gel electrophoresis unit, ultraviolet transilluminator, and system for documenting gel image). Depending on the design and number of primer mixes prepared, PCR-SSP can be used to provide a low- to medium-resolution type for each locus. If the low resolution type is known, additional PCR-SSP testing

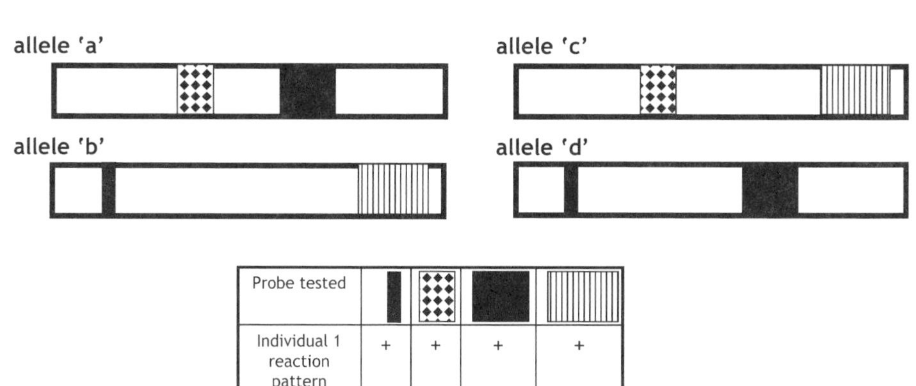

Probe tested				
Individual 1 reaction pattern	+	+	+	+
Individual 2 reaction pattern	+	+	+	+

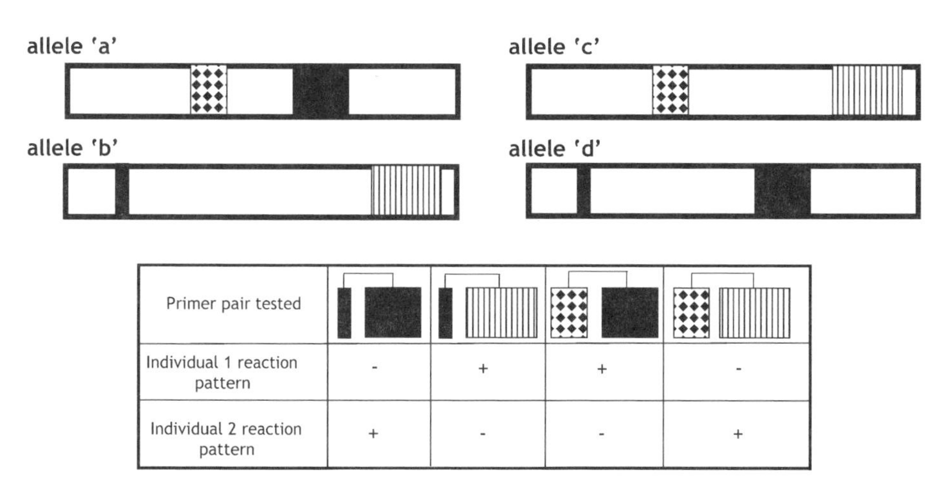

Primer pair tested				
Individual 1 reaction pattern	-	+	+	-
Individual 2 reaction pattern	+	-	-	+

Fig. 2. Diagrammatic representation of ambiguous human leukocyte antigen (HLA) types. The polymorphic HLA alleles from two unrelated individuals are represented as boxes, with the sequence motifs that distinguish the alleles given as shaded boxes. Both individuals are heterozygous and do not share either of their two HLA alleles. Using probes that target each motif individually **(A)**, both individuals will have the same probe reaction pattern with the combination of alleles "a" and "b" not being distinguished from "c" and "d." However, use of PCR primers that link *cis* polymorphisms **(B)** allows differentiation of the *cis* and *trans* sequence motifs.

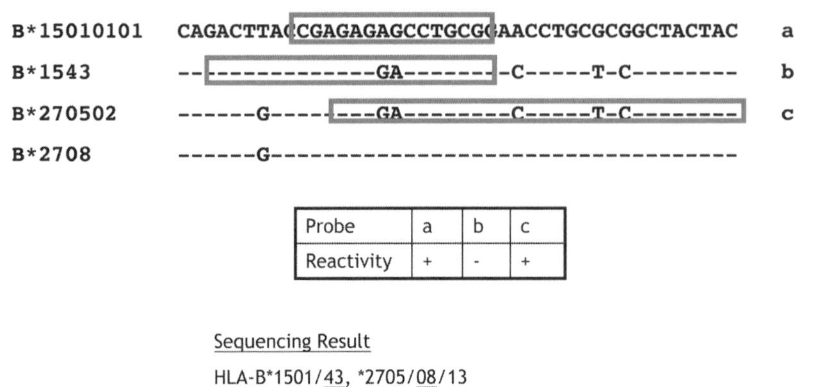

Probe	a	b	c
Reactivity	+	-	+

Sequencing Result

HLA-B*1501/43, *2705/08/13

PCR-SSO Result

HLA-B*1501/01N/26N/33/34/60, *2703/05/13/17/19

Fig. 3. Use of oligonucleotide probes to link *cis* polymorphisms. In this example, the presence of DRB1*1543 and B*0708 have not been excluded by PCR sequencing. However, probes b and c allow linkage of polymorphisms specific to B*1543 (probe b) and shared by B*1543 and B*270502, but not B*2708 (probe c). As probe b is negative, the reactivity of probe c is assigned to B*270502.

can be used to upgrade the low resolution type to a high resolution and/or allele level type **(Fig. 4B)**. It is also possible to go direct to high-resolution result using an extended panel of primer mixes. Such methods have been described in the literature and are now available commercially.

PCR-SSP is currently the quickest of the DNA-based methodologies and therefore can be used to provide fast HLA typing for cadaveric donor solid organ transplantation. The cost of PCR-SSP can be kept low with in-house developed primer mixes used, but this requires the laboratory to ensure that allele lists and interpretation tables are kept up-to-date. The laboratory must also have the ability to search primer sequences against current alleles as more alleles are defined.

PCR-SSP is not used for large-scale HLA typing such as that undertaken by the larger haematopoietic stem cell registries because of the rate-limiting step of agarose gel electropheresis. This step could be improved by the use of fluorescence detection systems such as that used in real-time PCR, which allows both detection and quantification of PCR amplicons and may allow discrimination of homozygous and heterozygous polymorphisms. Real-time PCR is also more

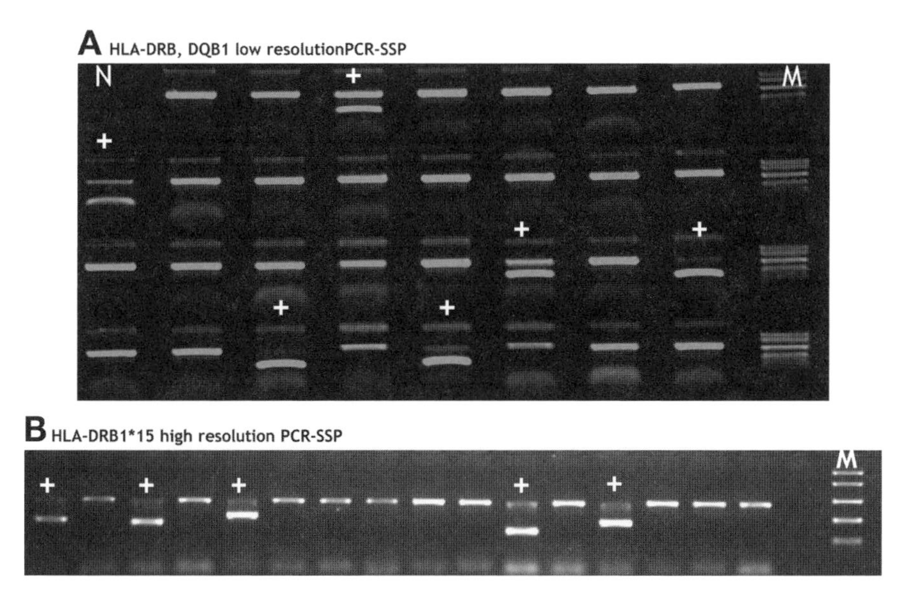

Fig. 4. PCR using sequence-specific primers (PCR-SSP) results for low- and high-resolution typing. N = negative control; M = molecular weight marker; + = positive reaction. The band present in all lanes except the negative control is the positive internal control amplicon.

An example of a low-resolution human leukocyte antigen (HLA)-DRB, DQB1 PCR-SSP typing result is given in (**A**). All HLA-DRB1, 3, 4, 5, and DQB1 specificities are covered by this test, which uses 31 primer pairs. The positive reactions in lanes 4, 9, 22, 24, 27, and 29 (numbering of lanes is from top left hand corner, reading from left hand side to right hand side) indicates the presence of HLA-DRB1*04, *15; DRB4*01; DRB5*01, and DQB1*03, *05. The results of further PCR-SSP testing of the same sample for high resolution (allele level) HLA typing of the DRB1*15 allele only are given in (**B**). Using 16 primer pairs, the four-digit alleles of the DRB1*15/*16 group can be distinguished. Positivity in lanes 1, 3, 5, 11, and 13 indicate the presence of DRB1*1501.

open to automation as a result of the nonrequirement for electrophoresis. Simpler systems involving the addition of PicoGreen® dye, which binds to the minor groove of double-stranded DNA molecules produced during the extension phases of the PCR reaction, can also be used. As the quantity of double-stranded DNA increases, the fluorescent emissions from the PicoGreen also increase, and this can be detected using a fluorescence reader. Such fluorescent systems, however, do not directly allow discrimination of molecular weights, and therefore false-negative results may be missed.

One of the advantages of PCR-SSP is the reduction in the number of ambiguous results compared with PCR-SSO (*see* **Fig. 2B**). *Cis*-polymorphisms can be linked by the generation of a product using primers designed to amplify from sequences containing both mutations.

2.3. PCR Sequencing

The resolution of results obtained with both PCR-SSO and PCR-SSP is limited by the number of oligonucleotide probes and primer mixes used, respectively. Current protocols do not cover the full region of the polymorphic exons targeted, and therefore it is always possible that novel mutations may not be observed. Nucleotide sequencing of HLA genes should, in principle, give the highest possible resolution results, allowing identification of all known polymorphisms and any novel polymorphisms. HLA gene sequencing was originally described using cloned PCR products to allow unambiguous identification of allele sequences, and the data generated from these studies led to the establishment of sequence databases that are used to develop reagents used in methods such as PCR-SSO and PCR-SSP. Improvements in automated DNA sequencing in the 1990s, concomitant with the introduction of robust dye-terminator chemistry protocols to allow accurate distinction between homozygous and heterozygous sequences, has permitted the establishment of routine HLA sequencing protocols from PCR-produced amplicons. Current PCR sequencing protocols have been described for all HLA loci using generic PCR amplification systems, and also group specific PCR sequencing protocols are described ***(12–14)***. The latter approach has been most useful for analysis of the HLA-DRB1 locus, as it is difficult to amplify all HLA-DRB1 alleles in a single PCR reaction without coamplification of other HLA-DRB alleles as a result of sequence similarities ***(15)***.

Sequencing heterozygous PCR products necessitates the use of dedicated software to allow identification of all heterozygous positions, and to allow comparison of the various possible sequences obtained against a database of known HLA sequences ***(16,17)***.

3. HLA Typing Resolution Requirements for Hematopoietic Stem Cell Transplantation

Usually, the search for a hematopoietic stem cell donor begins within the patient's family, and the identification of a matched related donor can be made with a minimum HLA-A, B, DRB1 type when the haplotypes present are unique. However, when similar haplotypes exist (e.g., if one of the parents is apparently homozygous by low resolution typing [*see* **Fig. 5** for an example]), it is advisable to perform more extended typing at higher resolution within families, and to include HLA-C and DQB1. HLA-DPB1 typing is also informative

	MOTHER		FATHER	
	a	b	c	d
HLA-A	*0101	*1101	*0201	*0201
HLA-C	*0701	*0601	*1203	*0701
HLA-B	*0801	*5701	*3901	*3906
HLA-DRB1	*0301	*0701	*0401	*0402
HLA-DRB3 (DR52)	*0101			
HLA-DRB4 (DR53)		*0103N	*0101	*0101
HLA-DRB5 (DR51)				
HLA-DQB1	*0201	*0303	*0301	*0302
HLA-DPB1	*0401	*0201	*0401	*1001

	PATIENT		SIBLING 1		SIBLING 2	
	a	c	a	d	a	c
HLA-A	*0101	*0201	*0101	*0201	*0101	*0201
HLA-C	*0701	*1203	*0701	*0701	*0701	*1203
HLA-B	*0801	*3901	*0801	*3906	*0801	*3901
HLA-DRB1	*0301	*0401	*0301	*0402	*0301	*0401
HLA-DRB3 (DR52)	*0101		*0101		*0101	
HLA-DRB4 (DR53)		*0101		*0101		*0101
HLA-DRB5 (DR51)						
HLA-DQB1	*0201	*0301	*0201	*0302	*0201	*0301
HLA-DPB1	*0401	*0401	*0401	*1001	*0401	*0401

Fig. 5. Identification of a matched related donor may require extended human leukocyte antigen (HLA) typing. This example of HLA typing within a family, highlights the potential for error when only low resolution HLA-A, B, DRB1 typing is performed (underlined HLA types). At low-resolution typing, the father appears to be homozygous for HLA-A, B, and DRB1, and both siblings appear to match the patient. However, extending the HLA type to include HLA-C, DQB1, and DPB1 demonstrates the father to possess two different haplotypes, and only one of the two siblings is a true match for the patient.

as recombination between HLA-DPB1 and DQB1 has been observed in as many as 5% of related cases as a result of a lack of linkage disequilibrium between HLA-DPB1 and other HLA loci ***(18)***.

When a search is extended to unrelated donors, it is an increasing requirement to perform high resolution or allele level typing for both patient and selected unrelated donor. The largest volunteer donor registry, the National Marrow Donor Program (NMDP), since June 2005 requires high resolution typing of all unrelated donors and patients for HLA-A, B, C, and DRB1 and it is likely that this requirement will extend to other registries ***(19)***. There are now extensive data supporting the importance of matching at high resolution for HLA-A, B, C, and DRB1 loci vs low resolution ***(20)***. There are also data to support the use of single allele mismatched donors when a perfect match cannot be found *(21)*. In order to know if a donor is only mismatched for one allele, high resolution is required to eliminate mismatches not detected by lower resolution typing methods. The role of HLA-DQB1 and DPB1 matching has been more difficult to ascertain. As a result of linkage disequilibrium between HLA-DRB and DQB1 loci, there are few cases of donor and patient pairs mismatched for HLA-DQB1 in the presence of no other mismatches; however, the presence of DQB1 mismatches in addition to other HLA mismatches was associated with increased mortality in a study published by the Fred Hutchinson Cancer Research Center *(21)*. Yet, the NMDP did not find a negative contribution of HLA-DQB1 mismatching in their analysis *(20)*. The differences in these studies is likely a reflection of differences in the cohorts used and analyses performed. As HLA-DPB1 is not in linkage disequilibrium with other HLA loci, most unrelated donors are HLA-DPB1-mismatched, so the role of HLA-DPB1 matching vs mismatching (on an HLA-A, B, C, DRB, DQB1 background) has been challenging to address. Data from the Anthony Nolan Trust and the International Histocompatibility Workshop support the function of HLA-DPB1 in allo-recognition, with HLA-DPB1 matching being associated with an increased risk of disease relapse *(22)*.

4. Patient and Donor Testing

Selecting a method for patient and donor testing depends on various factors including the number of samples that the laboratory tests, the type of sample obtained, and also the level of resolution to be achieved.

If a laboratory receives few samples in any working week, then the methodology of choice is most likely to be PCR-SSP, which can provide HLA types at various levels of resolution. The patient and related donor may initially be tested to low/medium resolution (PCR-SSO or PCR-SSP) with additional high/allele level resolution (PCR-SSP and/or PCR-Sequencing) performed if required to resolve ambiguities within family, or if an unrelated donor search is initiated. All potentially matching unrelated donors selected from the various volunteer donor registries must have their HLA type confirmed by repeat testing and this should always include extension of the original HLA type to include additional loci: e.g., if donor is HLA-A, B, and DRB typed on the Register, further testing for HLA-C, DQB1, and DPB1 is warranted, and loci (HLA-A, B, C, and DRB1 as a minimum) should be tested to high/allele level of resolution (PCR-SSP and/or PCR-sequencing) (*see* **Table 2**).

HLA genes are very polymorphic and determination of accurate HLA types can be complex and may necessitate the use of more than a single method. The development of DNA based HLA typing assays has made reliable, reproducible, and sensitive assays available for all laboratories supporting a haematopoietic stem cell transplantation program.

Table 2
Levels of Human Leukocyte Antigen Typing Resolution Obtained Based on Current Methods Most Widely in Use (Four-Digit Nomenclature Used)

	Example of resolution	PCR-SSP	PCR-SSO	PCR sequencing
Low	A1, 3 or A*01, *03	Yes	Yes	Yes (if full interpretation not performed)
Medium	A*0101/0102/0104N/0106/0109, *0301/0301N/0303N/0304/0311N/0313/0314	Yes	Yes	Yes (if full interpretation not performed)
High	A*0101/0104N, *0301/0301N	Yes (usually subtyping method)	Yes (less frequently)	Yes
Allele	A*0101, *0301	Yes (usually subtyping method)	Occasionally	Yes (may also require subgrouping PCR)

PCR-SSP, PCR using sequence-specific primers; PCR-SSO, PCR using sequence-specific oligonucleotides.

Acknowledgments

Dr. Raymond Fernando and Mr. Franco Tavarozzi are thanked for providing Fig. 4.

References

1. Warrens, A. and Lechler, R. (2000) HLA in health and disease (Lechler, R. and Warrens, A., eds.). Academic, London, UK, pp. 139–146
2. Mallal, S., Nolan, D., Witt, C., et al. (2002) Association between presence of HLA-B*5701, HLA-DR7, and HLA-DQ3 and hypersensitivity to HIV-1 reverse-transcriptase inhibitor abacavir. *Lancet* **359,** 727–732.
3. Belich, M. P., Madrigal, J. A., Hildebrand, W. H., et al. (1992) Unusual HLA-B alleles in two tribes of Brazilian Indians. *Nature* **357,** 326–329.
4. Klein, K. (1986) *Natural History of the Major Histocompatibility Complex*. John Wiley and Sons, Inc., Toronto, Canada.
5. Cao, K., Chopek, M., and Fernandez-Vina, M. A. (1999) High and intermediate resolution DNA typing systems for class I HLA-A, B, C genes by hybridization with sequence-specific oligonucleotide probes (SSOP). *Rev. Immunogenet.* **1,** 177–208.
6. Balazs, I., Beekman, J., Neuweiler, J., Liu, H., Watson, E., and Ray, B. (2001) Molecular typing of HLA-A, -B, and DRB using a high throughput micro array format. *Hum. Immunol.* **62,** 850–857.
7. Consolandi, C., Frosini, A., Pera, C., et al. (2004) Polymorphism analysis within the HLA-A locus by universal oligonucleotide array. *Hum. Mutat.* **24,** 428–434.
8. Li, A. X., Seul, M., Cicciarelli, J., Yang, J. C., and Iwaki, Y. (2004) Multiplexed analysis of polymorphisms in the HLA gene complex using bead array chips. *Tissue Antigens* **63,** 518–528.
9. Olerup, O. and Zetterquist, H. (1992) HLA-DR typing by PCR amplification with sequence-specific primers (PCR-SSP) in 2 hours: an alternative to serological DR typing in clinical practice including donor-recipient matching in cadaveric transplantation. *Tissue Antigens* **39,** 225–235.
10. Bunce, M., O'Neill, C. M., Barnardo, M. C., et al. (1995) Phototyping: comprehensive DNA typing for HLA-A, B, C, DRB1, DRB3, DRB4, DRB5 & DQB1 by PCR with 144 primer mixes utilizing sequence-specific primers (PCR-SSP). *Tissue Antigens* **46,** 355–367.
11. Newton, C. R., Graham, A., Heptinstall, L. E., et al. (1989) Analysis of any point mutation in DNA. The amplification refractory mutation system (ARMS). *Nucleic Acids Res.* **17,** 2503–2516.
12. van der Vlies, S. A., Voorter, C. E., and van den Berg-Loonen, E. M. (1998) A reliable and efficient high resolution typing method for HLA-C using sequence-based typing. *Tissue Antigens* **52,** 558–568.
13. Turner, S., Ellexson, M. E., Hickman, H. D., et al. (1998) Sequence-based typing provides a new look at HLA-C diversity. *J. Immunol.* **161,** 1406–1413.
14. Swelsen, W. T., Voorter, C. E., and van den Berg-Loonen, E. M. (2005) Sequence-based typing of the HLA-A10/A19 group and confirmation of a pseudogene coamplified with A*3401. *Hum. Immunol.* **66,** 535–542.

15. Kotsch, K., Wehling, J., and Blasczyk, R. (1999) Sequencing of HLA class II genes based on the conserved diversity of the non-coding regions: sequencing based typing of HLA-DRB genes. *Tissue Antigens* **53,** 486–497.
16. Sayer, D. C., Goodridge, D. M., and Christiansen, F. T. (2004) Assign 2.0: software for the analysis of Phred quality values for quality control of HLA sequencing-based typing. *Tissue Antigens* **64,** 556–565.
17. Rozemuller, E. H. and Tilanus, M. G. (2000) Bioinformatics: analysis of HLA sequence data. *Rev. Immunogenet.* **2,** 492–517.
18. Buchler, T., Gallardo, D., Rodriguez-Luaces, M., Pujal, J. M., and Granena, A. (2002) Frequency of HLA-DPB1 disparities detected by reference strand-mediated conformation analysis in HLA-A, -B, and -DRB1 matched siblings. *Hum Immunol.* **63,** 139–142.
19. Bioinformatics.NMDP.Org. Policies (www.nmdpresearch.org/HLA/hla_policies.html). Accessed June 2006.
20. Flomenberg, N., Baxter-Lowe, L. A., Confer, D., et al. (2004) Impact of HLA class I and class II high-resolution matching on outcomes of unrelated donor bone marrow transplantation: HLA-C mismatching is associated with a strong adverse effect on transplantation outcome. *Blood* **104,** 1923–1930.
21. Petersdorf, E. W., Anasetti, C., Martin, P. J., et al. (2004) Limits of HLA mismatching in unrelated hematopoietic cell transplantation. *Blood* **104,** 2976–2980.
22. Shaw, B. E., Potter, M. N., Mayor, N. P., et al. (2003) The degree of matching at HLA-DPB1 predicts for acute graft-versus-host disease and disease relapse following haematopoietic stem cell transplantation. *Bone Marrow Transplant* **31,** 1001–1008.
23. Robinson, J., Waller, M. J., Parham, P., et al. (2003) IMGT/HLA and IMGT/MHC: sequence databases for the study of the major histocompatibility complex. *Nucleic Acids Res.* **31,** 311–314.

4

Sequence-Specific Primed PCR (PCR-SSP) Typing of HLA Class I and Class II Alleles

Klara Dalva and Meral Beksac

Summary

For donor selection in hematopoetic stem cell transplantation, two-digit sequence-specific oligonucleotide (SSO) typing may be sufficient in the related sibling transplant setting. However, SSO typing is not sensitive enough to differentiate between the alleles that share the same *cis-trans* linkeage sequence casettes. In unrelated donor selection, PCR using sequence-specific primers, a flexible and widely used method known to cause less ambiguious results, may be preferable. However, this technique is limited by the number of the samples that can be processed at one time and also by the number of the primer mixes that can be utilized.

Key Words: PCR-SSP; HLA typing.

1. Introduction

HLA matching between the donor and recipient improves the success of unrealted hematopoietic cell transplantation *(1,2)*. The use of PCR-SSP for human leukocyte antigen (HLA) typing was first described for Class II and later for Class I specificities *(3,4)*. The principle of PCR using sequence-specific primers (PCR-SSP) is discussed detailed in Chapter 3. Briefly, this technique is based on the use of sequence-specific primers designed to be complementary or noncomplementary to the target DNA sequence. This approach is also called amplification refractory mutation system (ARMS) *(5)*. Specificity for PCR-SSP is derived from matching the terminal 3′-nucleotide of the primers with target DNA sequences. Primer template mismatches other than the 3′-mismatch (internal mismatches) may also effect the specificity of the primer *(6)*. The amplified DNA is usually visualized via agarose gel electrophoresis containing ethidium bromide. PCR-SSP protocols using fluorogenic probes have been reported as well *(7)*. Depending on the primer mixes used, it is possible to

From: *Methods in Molecular Medicine, vol. 134: Bone Marrow and Stem Cell Transplantation*
Edited by: M. Beksac © Humana Press Inc., Totowa, NJ

achieve low and/or high resolution typing (two to four digits) of HLA Class I and /or II alleles. There are various commercially available kits, and it is recommended to use kits that incorporate positive control primers pairs which amplify a segment of a non-HLA region (e.g., human growth hormone). Here, we describe one of the commercially available techniques developed by Olerup (Genovision As, Oslo, Norway).

2. Materials

2.1. Sample

DNA is usually obtained from EDTA or ACD anticoagulated venous blood samples. Heparin interferes with Taq polymerase activity, and should not be used as an anticoagulant. By using the GenoPrep™ B350 method, which is described in this chapter, DNA extraction can also be performed with heparinized samples. If the white blood cell (WBC) counts are very low, the use of buffy coat is suggested. DNA can also be extracted from fresh or frozen blood samples. Storage of samples for less than 5 d at +4°C is recommended. Bone marrow aspiration material is also acceptable

2.2. DNA Extraction

2.2.1. GenoM™ 6 Robotic Workstation

GenoPrep B350 cartridges accompanying accesories (the kit includes prefilled reagent cartridges, 2-mL screw-cap Eppendorf tubes required for distribution of blood samples, 1.5-mL screw-cap Eppendorf tubes for the elution of extracted DNA, plugged pipet tips, and pipet tip holders sufficient for DNA isolation from 48 samples).

2.2.2. Salting-Out Technique

This method is a modification of Miller's salting-out procedure, with the ommision of proteinase K and addition of a chloroform extraction phase *(4)*.

1. Red cell lysis buffer (RCLB): 0.144 *M* ammonium chloride (NH_4Cl), 1 m*M* sodium bicarbonate ($NaHCO_3$): 15.4 g of NH_4Cl and 1.68 g of $NaHCO_3$ are dissolved in 2 L of double-distilled (dd)H_2O.
2. Nuclear lysis buffer (NLB): 10 m*M* Tris-HCl pH 8.2, 0.4 *M* sodium chloride (NaCl), 2 m*M* disodium EDTA (Na_2EDTA) pH 8.0. Dissolve 23.37g of NaCl in 900 mL of dH_2O. Add 10 mL 1 *M* Tris-HCl pH 8.2 and 10 mL Na_2EDTA pH 8.0 and adjust to 1 L with ddH_2O.
3. 10% w/v sodium dodecyl sulfate (SDS): dissolve 100 g of SDS in 1 L ddH_2O. Store at room temperature to prevent the formation of precipitates.
4. NLB + SDS buffer: 300 mL of NLB and 20 mL of 10 w/v SDS are mixed. Store at room temperature.

5. 95% Ethanol. Dilute 950 mL of absolute ethanol with 50 mL of ddH_2O.
6. 70% Ethanol. Dilute 700 mL of absolute ethanol with 300 mL of ddH_2O.
7. 6 *M* NaCl: 350.64 g sodium chloride is dissolved in 800 mL ddH_2O by warming the solution followed by adjusting the volume to 1 L. As 6 *M* NaCl is a saturated solution, not all of the NaCl will dissolve into the solution.
8. Centrifuge.
9. Plastic tubes (15 mL, 2 mL).

2.3. PCR Amplification

1. SSP™ HLA-ABC low resolution kit, unlicensed (the licensed kits contain Taq polymerase within the PCR Master Mix). Kit content:
   a. 96-Well PCR plates, which are precoated with the primers (wells 1–23 HLA-A, wells 25–71 HLA-B, wells 73–95 HLA-C, wells 24, 72, and 96 negative control primers), and are covered with PCR compatible aluminium foil.
   b. PCR master mixes: the final concentration (fc) of each dNTP is 200 μ*M*, PCR buffer (KCl fc: 50 m*M*, $MgCl_2$ fc: 1.5 m*M*, Tris-HCl, pH 8.3 fc: 10 μ*M*, 0.001 w/v gelatin, glycerol fc: 5%, cresol red fc: 100 μg/mL).
   c. PCR caps.
2. Olerup SSP HLA-DQ-DR SSP combi tray, unlicensed (the licensed kits contain Taq polymerase within the PCR Master Mix). Kit content:
   a. 32-Well PCR plates, which include the primers (wells 1–8 DQ, wells 9–31 DR primers, well 32 negative control) covered with PCR compatible aluminium foil.
   b. PCR master mix contents are similar to that described in **item 1**.
   c. PCR caps.

   Kits maintain their stability for 21 mo at –20°C.
3. Taq polymerase (5 U/μL).
4. PCR thermocycler(s): model with a 96-well format (e.g., Corbett Research CGI-960, Perkin Elmer 9600 or 9700, MJResearch PTC200 PCH).
5. Dispensing equipment: adjustable pipet(s), electronic multistep dispenser, plastic tips.

2.4. Agarose Gel Electrophoresis

1. 10X TBE buffer (1 L, pH: 8.0):108 g Tris base; 55 g boric acid; 14.8 g EDTA (sodium salt). Dissolve all ingredients in ddH_2O and adjust the volume to 1 L. The solution is filtered to remove particles. Store at room temperature.
2. 0.5X TBE buffer (1 L): dilute 50 mL of 10X TBE buffer to 1 L with ddH_2O. Store at room temperature.
3. 2% (w/v) Agarose: (150 mL): 3 g of agarose dissolved in 150 mL 0.5X TBE buffer by heating in a microwave oven (approx 2 min at 800 W). Once the gel has cooled to 60°C, the gel is then stained with ethidium bromide (0.625 mg/mL) by adding three drops of EtBr to 150 mL of gel. Agarose solutions can be stored at 50°C for up to 1 wk.

4. Ethidium bromide solution (10 mL, 0.625 mg/mL): dissolve 0.625 g of EtBr in 10 mL ddH_2O and stir for several hours until dye has dissolved completely. Store in ambient conditions for up to 1 yr in an amber bottle or protected from light with aluminium foil.
5. Molecular weight marker: 100-bp ladder.
6. Eight-channel electronic multidispenser.
7. Electrophoresis tank.
8. Horizontal gel casting forms with combs (1 mm × 17 well) suitable for multi-channel loading.
9. Power supply.
10. Ultraviolet transilluminator.

2.5. Documentation

Digital imaging system (Dolphine imaging system).

2.6. Interpretation

1. *Manual:* Refer to the interpretation and specificity tables provided with the typing kits.
2. *Computer-based:* The Score™ interpretation software is recommended.

3. Methods

3.1. DNA Extraction

Extracted, highly pure DNA is required for PCR-SSP typing. Because the class I amplicon is larger than the class II amplicon, the DNA preparation method is critical for the success of the PCR. Here, two different methods are described. The optimal DNA concentration for GenoM 6 isolations is 15 μg/mL and 30 μg/mL for the other methods. High-quality DNA results in an optimal density 260/280 ratio greater than 1.6 (*see* **Notes 1–3**).

3.1.1. GenoM 6 Robotic Workstation

The GenoM 6 Robotic Workstation can be used for an easy, automated nucleic acid isolation and purification. The procedure involves binding of Genoprep magnetic beads to the nucleic acids. The GenoM 6 Robotic Workstation performs all the steps involved in sample preparation, including sample lysis, binding of nucleic acids to the beads, and washing and elution in an automated system. Protocols can be scaled up and down according to the needs of the user.

3.1.1.1. Principle

Isolation of the DNA relies upon it's binding to the silica surface of the paramagnetic beads in the presence of a chaotropic salt solution such as sodium iodide, guanidium thiocyanate, or guanidium hydrochloride.

3.1.1.2. Isolation Steps

1. Lysis and binding: addition of chaotropic solutions (GTC) and silica magnetic beads results in complete cell lysis and DNA release. DNAses are denaturated and inactivated during this step. Magnetic beads are mixed with the sample; resulting in binding of the DNA to the GenoPrep DNA magnetic beads. Subsequently, immobilized DNA bound to magnetic beads, are collected by application of a magnetic force.
2. Washing steps: three consecutive washing steps are performed during the application of magnetic force. During the first wash, a chaotropic solution removes any unbound material. In the second wash, ethanol is applied and removes GTC, followed by a third wash, in which water removes the ethanol.
3. Elution: the DNA is released from the GenoPrep DNA magnetic beads via thorough mixing with added water

3.1.1.3. Starting Material

Begin with 350 μL of a homogenized anticoagulated (treated with EDTA, ACD or heparin) blood sample.

3.1.1.4. Operation Description

1. Select the protocol card (IC card) specified for DNA isolation and insert it into the IC card inlet.
2. Switch on the main power of the GenoM 6 Instrument.
3. Choose the desired volume from the top menu. 350 μL protocol is selected for a low resolution typing of HLA-A, -B, -C, -DR, -DQ.
4. Load disposables on the instrument platform (1.5-mL elution tubes to the first row, tips to the second row, and 2-mL sample tubes to the third row. Prefilled cartridges are inserted in the predefined positions.
5. Place the sample tubes that include a minumum of 1 mL of a homogenized anticoagulated blood sample on the instrument platform.
6. Close the instrument and start the extraction procedure. The Menu Screen will indicate at which stage the isolation process is in.
7. Retrieve the isolated DNA from the elution (first) row.
8. Proceed with the PCR amplification step immediately or, otherwise store the DNA at +4°C for up to 7 d or at –20°C for several months.

3.1.2. Salting-Out Technique

1. Centrifuge 5 mL of EDTA or ACD-A anticoagulated blood to obtain a buffy coat.
2. Aspirate the buffy coat into a 15-mL polypropylene tube.
3. Add 10 mL of RCLB, invert several times, and leave to stand for 5 min at room temperature.
4. Centrifuge the tube at 1000*g* for 10 min. Discard the supernatant and gently rinse the pellet with 2 mL of RCLB buffer. The pellet will appear white with a pink

halo. If there is excess haemoglobin, resuspend the pellet in RCLB and gently agitate, continuing with further centrifugation (10 min,1000*g*).

5. Resuspend the pellet in 3 mL of NLB + SDS buffer. Add 1 mL of 6 *M* NaCl and vortex. At this stage, the precipitate should be visible.
6. Add 2 mL of chloroform and gently mix until a homogenous milky solution is obtained. Centrifuge the tube (10 min, 1000*g*).
7. Aspirate the top phase containing the DNA into a 20-mL tube. Avoid aspirating protein from the interphase. If the DNA phase is not clear, transfer the aspirate into a clean polypropylene tube and repeat the chloroform extraction step.
8. Add 2 mL of 95% ethanol and gently invert the tubes until all of the DNA is precipitated.
9. Centrifuge the tubes (5 min, 700*g*) and resuspend the pellet in 2 mL of 70% ethanol. Repeat again.
10. Transfer the DNA precipitate into a sterile 0.5-mL microcentrifuge tube and centrifuge the tubes in order to obtain a DNA pellet. Remove the excess ethanol by a further centrifugation step.
11. Resuspend the DNA in 300 μL of sterile ddH_2O. A DNA yield of 0.2–1.0 mg/mL is expected from 5 mL of blood.

3.2. PCR Amplification

The primer set from HLA-ABC low resolution kit contains 5′ and 3′ primers for grouping HLA-A*0101 to -A*8001, -B*0702 to -B*8301, -Cw*0102 to -Cw*1802, and for the recognition of Bw4 & Bw6 sequence motifs. The HLA-DQ-DR SSP combi tray contains 5′ and 3′ primers, required for grouping the DQB1 alleles into the serological groups DQ2 to DQ9, serological groups DR1 to DR18. Furthermore primer pairs that recognize the DRB3, DRB4, and DRB5 groups of alleles are provided. These products utilise the ARMS technology.

3.2.1. HLA-ABC Low-Resolution Typing

For HLA-ABC typing of one plate:

1. 12 μL PCR Mastermix and 8.3 μL Taq polymerase are added to a 0.5-mL Eppendorf tube.
2. Dispense 3 μL of the Mastermix-Taq Polymerase mixture to wells with numbers 24, 72, and 96.
3. Add 7 μL of dH_2O to these three wells, which contain primer mixes for the negative control.
4. Add 202 μL of DNA (30 ng/μL), and 505 μL dH_2O to the remaining Master mix-Taq polymerase mixture (311.3 μL) and mix throughly.
5. Dispense 10 μL of the final mixture into wells 1–23, 25–71, and 73–95.
6. Seal the plates with the lids provided (*see* **Note 4**).
7. Proceed with amplification in a thermal cycler equipped with a heated lid. The temperature gradient across the heating block should be <1°C.

3.2.2. HLA-DR-DQ Low-Resolution Typing

For HLA- DQ-DR typing of one plate:

1. Add 114 µL of PCR Master mix and 3 µL Taq polymerase to a 0.5-mL Eppendorf tube. Mix throughly.
2. Dispense 3 µL of the Mastermix-Taq Polymerase mixture to well no. 32, which contains the negative primer mix.
3. Add 7 µL of dH_2O to well no. 32.
4. Add 74 µL of DNA (30 ng/µL) and 182 µL of dH_2O to the remaining Master mix-Taq Polymerase mixture (114 µL).
5. Mix throughly.
6. Dispense 10 µL of this final mixture into wells 1–31.
7. Seal the plates with the lids provided (*see* **Note 4**).
8. Proceed with amplification in a 96-well thermal cycler equipped with a heated lid. The temperature gradient across the heating block should be <1°C.

3.2.3. PCR Cycling Condition

Identical PCR cycling conditions are used for all applications (*see* **Notes 5–8**):

1 cycle	94°C	2 min	denaturation
10 cycles	94°C	10 s	denaturation
	65°C	60 s	annealing and extention
20 cycles	94°C	10 s	denaturation
	61°C	50 s	annealing
	72°C	30 s	extention

3.3. Agarose Gel Electrophoresis

1. Prepare 150 mL of 2% agarose in 0.5X TBE buffer and stain with EtBr (*see* **Note 9**).
2. Pour the agarose into the gel tray and insert the combs (17 wells per row, 6 rows)
3. Allow to stand for 30 min at room temperature.
4. Fill the electrophoresis tank with 550 mL of 0.5X TBE buffer. Gel must be submerged at least in 5 mm buffer. The buffer can be reused up to 15 times (*see* **Note 10**).
5. Once the gel is set, remove any sealing blocks or tape and and submerge the tray into the tank.
6. Using an eight-channel pipet, load the PCR products into the chambers of the gel.
7. For each row, load a DNA marker in one well.
8. Run the gel for 15–20 min at 10 V/cm.

3.4. Photography

1. Turn on the imaging system (Wealtec Dolphine-View digital imaging system).
2. Once the screen lights is on press "Start."
3. Insert a disk into the drive.

4. Open the door of the ultraviolet transilluminator and remove the mobile table.
5. Transfer the gel to the mobile table and take care to not create air bubbles.
6. Adjust the image using the leveling button present on the front panel.
7. "Freeze" the image before saving the file.
8. "Save" the image to the disk (maximum three images per disk).
9. The system automatically gives a tag number to every gel.
10. Turn back to the original view by pressing "Live."
11. Remove the floppy disk.
12. Remove and discard the gel accordance in to the laboratory safety rules.
13. Turn off the imaging system.
14. Save the gels in a new file in a computer, in order to analyse via the "Dolphin-View Band Tool software."
15. Open the analysis program and select the file to be analyzed.
16. Adjust the image to optimize the view of specific/control bands.
17. Record patient/gel information on the image and save the image as a new file.

3.5. Interpretation of Results

The HLA-ABC kit amplifies all of the HLA-A, -B, and -C alleles with a few exceptions. The HLA-DQB1 alleles listed in the 2005 Nomenclature are all amplified with this kit. This is also valid for HLA-DRB alleles which are amplified by this kit.

Alleles are assigned by the presence of specific PCR product(s). The size of each PCR product may be helpful in the interpretation of the results (*see* **Note 11**). Each PCR-SSP reaction is deemed to have worked if the internal control amplification is observed (*see* **Note 2**). In the presence of a specific amplification, the intensity of the control band often decreases (sometimes regarded as negative). Reactions without any control or allele specific amplicons are recorded as "not tested" (*see* **Notes 1**, **5**, and **6**). Nonspecific amplification, especially with GC-rich primers, may sometimes be observed (*see* **Note 3**). Because many alleles are amplified in more than one reaction, sporadic PCR failures usually do not affect the full assignment of the genotype.

Genotypes are assigned by identification of the pattern of positive and negative reactions and interpretation of these results according to the specificity tables present in the typing kits (*see* **Note 12**). The evaluation is suggested to be done by use of the Score interpretation software, which enables faster and more accurate interpretation.

4. Notes

1. Poor-quality or an insufficient amount of DNA may lead to reaction failures. The DNA quality should be checked. WBC counts less than 1000/μL may lead to an insufficient amount of DNA. Repeat the isolation with buffy coat collected from 10 mL instead of 350 μL of blood.

The DNA extracted by the GenoPrep kits is usually highly purified. If a DNA sample was isolated with a different extraction method, repeat the extraction with the GenoPrep kits by using diluted DNA which is adjusted to 350 μL with ddH_2O.

If degradation or poor-quality DNA is suspected, increase the amount of Taq (50% more) and reduce DNA volume.

2. Degraded DNA may fail to produce larger amplicons, such as those of control. In the presence of specific amplification, the intensity of the control bands often diminishes but do not disappear completely.
3. DNA concentration exceeding 50 ng/μL will increase the risk of nonspecific amplifications.
4. Failure to close the lids/bands properly, or insufficient pressure from above, may lead to evaporation during PCR reaction.
5. PCR program temperature errors may result in the absence of specific amplicons while the control bands exist.
6. Poor contact of the tubes with PCR block may cause (1) individual reaction failures or (2) failure in a part of reaction. Be sure to apply sufficient pressure from above and try to dip the tubes in light paraffin oil to coat the exterior of the tubes.
7. Interruption and restarting of PCR programs especially during the early cycles will lead to multiple bands (allele-specific bands) because of the induction of low stringency PCR. Use a continuous power supply and check the PCR machine for error messages.
8. Partial typing failures are a common problem of overused PCR thermocyclers.
9. Uneven distrubition of EtBr on the gel may lead to bright spots over the gel.
10. Wrong buffer composition, overuse of TBE buffer, and overheated buffer as a result of high voltage application may lead to blurred bands. Check and lower the voltage during gel electrophoresis and apply fresh buffer to the tank.
11. PCR fragments shorter than 125 bp have a lower intensity.
12. Unexpected reaction patterns may be a result of incorrect loading of the amplicons on the gel. Repeat the test procedure.

References

1. Petersdorf, E. W., Anasetti, C., Martin, P. J., et al. (2004) Limits of HLA mismatching in unrelated hematopoetic cell transplantation *Blood* **104,** 2976–2980.
2. Flomenberg, N., Baxter-Lowe, L. A., Confer, D., et al. (2004) Impact of HLA class I and class II high-resolution matching on outcomes of unrelated donor bone marrow transplantation: HLA-C mismatching is associated with a strong adverse effect on transplantation outcome *Blood* 104**,** 1923–1930.
3. Browning, M. J., Krausa, P., Rowan, A., Bicknell, D. C., Bodmer, J. G., and Bodmer, W. F. (1993) Tissue typing the HLA-A locus from genomic DNA by sequence-specific PCR: comparison of HLA genotype and surface expression on colorectal tumor cell lines. *Proc. Natl. Acad. Sci. USA* **90,** 2842–2845.
4. Olerup, O. and Zetterquist, H. (1992) HLA-DRB1*01 subtyping by allele specific pcr amplification: a sensitive, specific and rapid technique. *Tissue Antigens* **37,** 197–204.

5. Newton, C. R., Graham, A., Heptinstall, L. E., et al. (1989) Analysis of any point mutation in DNA. The amplification refractory mutation system (ARMS). *Nucleic Acids Res.* **17,** 2503–2516.
6. Bunce, M. and Welsh,K., (2000) PCR-SSP typing of HLA class I and class II alleles, in *ASHI Laboratory Manual,* 4th Ed. (Hahn, A. B., Land, G. A., and Strothman, R. M., eds.). ASHI Publications, pp: 1–19.
7. Blasczyk, R. (1998) New HLA typing methods, in *New Diagnostic Methods in Oncology and Hematology,* (Huhn, D., ed.), Springer, Berlin, Germany, pp. 143–195.

5

HLA Typing with Sequence-Specific Oligonucleotide Primed PCR (PCR-SSO) and Use of the Luminex™ Technology

Klara Dalva and Meral Beksac

Summary

The hybridization products obtained by PCR using sequence-specific oligonucleotides (PCR-SSO) can be traced either by colorimetric- (streptavidin- biotin), X-ray- (digoxigenin-CSPD), or fluorescence- (FITC, PE) based detection systems. To achieve a faster, reliable, automated typing, microbead and fluorescence detection technology have been combined and introduced to this field (XMAP™ technology). For each locus, a maximum of 100 microspheres, which are recognizable by their specific color originating from two internal fluorescent dyes, are used. Each microsphere is coupled with a single probe that is capable of hybridizing with the biotin labeled complementary amplicon. Once hybridization occurs, it can be quantified via the fluorecence signal originating from fluorescently (Streptavidin-PE) labeled amplicons captured by the beads. Currently, there are two commercially available systems that differ in the scale of probes and the method of amplification or denaturation. One of these will be described in detail in this chapter.

Key Words: PCR-SSO; Luminex™; XMAP™ technology; HLA typing.

1. Introduction

The first applications of PCR using sequence-specific oligonucleotides (PCR-SSO) in the field of human leukocyte antigen (HLA) were started with the DQA1 locus and followed by class I and other class II specificities ***(1–8)***. In order to eliminate ambiguities, protocols that use PCR primers designed to amplify the entire hypervariable region of a particular HLA locus (group-specific primers) were introduced. These PCR products are incubated in the presence of a panel of labeled oligonucleotides designed for the detection of different polymorphic positions specific for an allele or allelic group ***(8)***. The

From: *Methods in Molecular Medicine, vol. 134: Bone Marrow and Stem Cell Transplantation*
Edited by: M. Beksac © Humana Press Inc., Totowa, NJ

primers for HLA-A, -B, and -C loci usually give a locus-specific product covering exons 2 and 3. Primer for HLA-DR gives a product from exon 2 ***(1,7)***. The introductory chapter on molecular HLA typing authored by Ann-Margaret Little (Chapter 3) summarizes the main features of PCR-SSO typing and compares it with the other methods. Here, we will describe one of the two automated systems using Luminex^{TM} technology, that is capable of typing of multiple inviduals simultaneously ***(9–11)***.

The xMAP^{TM} technology developed by Luminex (Austin, TX) is a microsphere-based, multiplexed, flow cytometric analysis system that makes it possible to combine hybridization with fluorescence detection ***(12,13)***. Classification of HLA alleles by Luminex system was first described by Fulton et al. ***(14)***. The xMAP technology employs simultaneous applications of up to 100 probes, which are already coated on microspheres. To be able to recognize these probes, 100 shades of two colors that are formed by the ratio of two internal fluorescent dyes are assigned for each probe. The fluoroanalyzer contains a red laser that excites the dyes in the microspheres and categorizes them based on their dye content. The microspheres are coated with carboxyl groups in order to achieve a covalent bridge between the oligonucleotides that contain terminal amino groups and the beads. Thus, the bound oligonucleotides also become color-coded. In addition to the red laser, the instrument contains a green laser that is used to quantify fluorescently (Streptavidin-PE) labeled amplicons captured by the beads. Each probe mix contains one ore more oligonucleotide(s) that react with all alleles within the locus of interest, which also serves as an internal control. They are used in the normalization of the values during the calculation of reaction patterns. If the minumum fluorescent intensity values defined for this control probes are not achieved, the sample test must be repeated.

All probe-coated beads can be applied within a well. This unique feature enables one locus typing of 96 samples or more loci on smaller number of samples. Currently, there are two commercially available kits. The probe number, specificity, and cutoff levels may vary between lots and commercial sources. A comparison of available probes presented by the two commercially avilable sources are presented in **Table 1**. In this chapter, we will focus on one of these (Tepnel Life Codes).

2. Materials

2.1. Specimen

DNA is usually obtained from EDTA or ACD anticoagulated venous blood samples. Heparin interferes with Taq polymerase activity and should not be used as an anticoagulant. Final DNA concentration must be 10-200 ng/μL.

Table 1
Number of Probes for Human Leukocyte Antigen (HLA) Typing (January 2006)

Locus	LabType® SSO	Tepnel Life Codes
A	63	64
B	100	89
C	56	51
DRB1	70	68
DQA1	11	NA
DQB1	37	33
DPB1	40	59
Bw4 supp.	15	NA
HLA-DRB(Generic, DR52)	NA	82
DRB3,4,5	29	NA
DQA1	11	NA

SSO, sequence-specific oligonucleotides; NA, not assigned.

2.2. DNA Extraction

DNA extraction is as described in Chapter 4.

2.3. PCR Amplification

1. Lifecodes HLA-SSO typing kits (HLA-A, -B, -C, -DRB, -DQB). Each kit contains the master mix including 10X PCR buffer, 10X dNTPs, locus-specific biotin-conjugated primers, and a fluorescently coded probe mix and a dilution solution (DS) necessary for the hybridization step.
2. Microcentrifuge.
3. Vortex mixer.
4. Recombinant Taq polymerase (5 U/μL).
5. Nuclease-free water.
6. Cooling block or melting ice.
7. DNase/RNase-free tubes (0.5 mL).
8. DNase/RNase-free thin-wall tubes (0.2 mL).
9. PCR thermocycler(s): model with a 96-well format (e.g., Corbett Research CGI-960, Perkin Elmer 9600 or 9700, MJResearch PTC200 PCH).
10. Dispensing equipment: adjustable pipet(s), plastic tips.

2.4. Hybridization

1. Lifecodes HLA-SSO typing kits (HLA-A, -B, -C, DRB, -DQB).
2. Streptavidin-PE (SA-PE) (1 mg/mL, Lifecodes, cat no. 628511).
3. Heating block.

Table 2
Reaction Components Necessary for Amplification (Except DNA)

Reagent	Amount sufficient for 1 reaction (Prepare for *n* + 1 reactions)
Master Mix (thawed to room temp.)	15.0 µL
ddH_2O	24.5 µL
Taq Polymerase (–20°C)	0.5 µL
Total	40.0 µL

4. Ultrasonic bath (Branson).
5. Vortex mixer with adjustable speed.
6. PCR plates (thin-wall, 96-well, 250 µL/well).
7. Polyethylene sealing tape.
8. Dispensing equipment: adjustable pipette(s), plastic tips.

2.5. Data Collection

1. Luminex^{100} Instrument and XY Platform.
2. Luminex ^{100} IS software, Quick Type for Lifematch 2.1 software.
3. Sheath fluid.
4. Daily maintenance reagents (70% ethanol, double-distilled [dd]H_2O).

2.6. Analysis

Quick Type for Lifematch 2.1 software.

3. Method

3.1. DNA Extraction

DNA extraction is as described in Chapter 4 (*see* **Notes 1** and **2**).

3.2. PCR Amplification

The Lifecodes kits utilize an assymetric PCR that increase the amount of one primer approx 10 times and allow to generate single-stranded DNA products in addition to the double-stranded products (*see* **Note 3**).

1. Stand the master mix to reach room temperature.
2. Gently vortex for 10 s and centrifuge briefly to ensure salts are in solution (*see* **Note 4**).
3. Prepare the amplification mix sufficient for *n* + 1 samples (**Table 2**) (*see* **Note 5**).
4. Pipet 10 µL of genomic DNA isolated by GenoPrep™ (approx 200 ng of DNA) into the PCR tubes.

5. Dispence 40 µL of the amplification mix into the PCR tubes containing the DNA sample.
6. Close the tubes tightly and proceed with amplification.

3.3. PCR Cycling Conditions

Identical PCR cycling conditions are used for all applications (*see* **Notes 6** and **7**).

1 cycle	94°C	2 min	denaturation
1 cycle	95°C	5 min	denaturation
8 cycles	95°C	30 s	denaturation
	60°C	45 s	annealing
	72°C	45 s	extention
32 cycles	95°C	30 s	denaturation
	63°C	45 s	annealing
	72°C	45 s	extension
1 cycle	72°C	15 min	extension

3.4. Hybridization

This technique excludes the prehybridization denaturation step. Single-stranded DNA is achieved by assymetric PCR and by the addition of a 5-min incubation at 97°C prior to the hybridization (*see* **Notes 8** and **9**).

Before starting with the hybridization turn on the Luminex^{100} instrument and XY platform to allow warming of the laser at least for 30 min prior to the analysis.

1. Warm probe mix in 57°C (55–60°C) heat block for 7 (5–10) min to obtain solubilization of the components thoroughly. Protect from light to avoid photobleaching, do not refreeze bead mixture after thawing, store at 2–8°C (*see* **Note 10**).
2. In order to suspend the probe carrier beads, sonicate briefly for 15 s and then vortex for another 15 s (*see* **Note 10**).
3. Pipet 5 µL of locus specific PCR product (amplicon) into a thermal cycler 96-well plate (**Fig. 1**).
4. Aliquot 15 µL of probe mix into each well. When aliquoting for more than 10 samples, gently vortex the probe mix for every set of 10 (*see* **Note 11**).
5. Seal plate with polyethylene sealing tape (*see* **Note 12**).
6. Hybridize the samples under the following conditions using a thermocycler:

   97°C for 5 min
   47°C for 30 min
   56°C for 10 min
   56°C hold

7. While hybridization is in progress, dilute Streptavidin PE solution within the DS (dilution ratio: 1/200 for one sample 0.85 µL of SA-PE and 170 µL of DS). Pref-

Samples	1	2	3	4	5	6	7	8	9	10	11	12
HLA-A ➔	▼	▼	▼	▼	▼	▼	▼	▼	▼	▼	▼	▼
HLA-B ➔												
HLA-C ➔												
HLA-DRB$_{(generic)}$ ➔												
HLA-DR52 ➔												
HLA-DQ ➔												

Fig. 1. Sample of hybridization data sheet prepared for this application.

erably prepare the solution for $n + 1$ samples. Because SA-PE is light-sensitive, keep the DS/SA-PE solution in the dark at room temperature. Do not reuse; discard any remaining solution (*see* **Note 13**).

8. While the tray is still on the thermocycler at the 56°C "hold" step, serially aliquot 170 µL of SA-PE/DS mixture to each well including sample(s). It is critical to dilute all of the samples within 5 min following the 10-min incubation at 56°C.
9. Remove the samples from the thermocycler and place in the Luminex^{100} instrument to analyze the samples.
10. Turn off the thermal cycler.

3.5. Analysis

Assay the samples immediatly using the prewarmed Luminex^{100} instrument. If immediate reading is not possible, protect samples from light. Samples can be read within 30 min following the SA-PE/DS dilution.

Prior to analysis, set up a "batch run" by which the samples will be analyzed.

3.6. Data Acquisition

The steps described next is a general guide for data acquisition. Daily start-up, calibration, maintenance, and shut-down procedures may be found in the User's Manual.

1. Turn on the Luminex^{100} 30 min to 4 h before acquisition of the samples.
2. Prior to acquisition, check the level of the sheath fluid, tighten the cap, empty the waste tank, perform a "prime" and a wash with 70% ethanol, and proceed with a wash using the sheath fluid.

3. Set up a "batch run" by which the samples will be analyzed. For each locus, follow the directions for the analysis.
4. Open Luminex[100] IS software from desktop and open "submitted batch." Follow directions. To analyze multiple loci in one plate, "Create Multi-batch" by adding batches and click finish after naming the Multi-Batch.
5. Eject the plate holder and place the 96-well plate containing the samples in the XYP heater block present on the plate holder. Click the "retract" icon to start the acquisition.
6. Once the sample tray has been placed into the XY Platform of the Luminex[100] instrument, click the "Start Plate" icon to start with acquisition.
7. After running of the samples, perform a sanitization wash by rinsing two times with 70% ethanol and perform "soak" with dH_2O.
8. Release the pressure on the sheath fluid tank.
9. Turn off the instrument. Completed batches are exported automatically as comma-separated values (csv) and are named as "output.csv" and saved in a folder with batch name defined on the third step. This data are available for the assignments of the results

3.7. Interpretation of Results

Interpretation of the results is performed by using the Quick Type for Lifematch 2.1 software with which the opened csv files may be analyzed.

The steps described next are only a general guide to data analysis.

1. Select "Final Typing Assignments."
2. Select file(s) to be analyzed.
3. Verify the number of counted events to be greater than 60 for each probe including the controls in each sample (this information presents in "data type" and count section of csv files).
4. The values for the concensus probes for each sample must exceed the minumum median fluorescent intensity (MFI) (*see* **Notes 4** and **7**). These values can be found on the threshold tables supplied with each kit or can be loaded to the analyzer by the technical services during updating process. The minumum thresholds are lot-specific.
5. For each probe, compare the normalized value with the threshold values supplied with the kits. The software makes this assignment automatically. However, the current result must be validated before continuing to the next step.

   Normalized values are calculated as:

$$\frac{\text{MFI (probe)} - \text{MFI (control blank or probe)}}{\text{MFI (concensus)} - \text{MFI (control blank or concensus)}}$$

6. If the measured value for a particular probe falls above the maximum threshold of a negative assignment and below the minumum value of a positive assignment, the software evaluates this value as "undetermined" (*see* **Note 14**).
7. Save and print the results.

4. Notes

1. Poor-quality, degraded, or insufficient amount of DNA may lead to reaction failures. Check DNA quality and purity. Reisolate DNA.
2. A sample contamination may lead to failure to yield an HLA typing result.
3. The differences between Lifecodes and LabType are: the amplification policy (asymetric PCR vs regular PCR) and the availability of ready to use mixtures and the amplification conditions; the number of probes are also currently different, and are summarized in **Table 1**.
4. Pipetting errors owing to poor homogenization of reagents (Master Mix) may result in reaction failures (low MFI for the control probes). This can be avoided by prewarming the Master Mix at 37°C for 5 min.
5. Poor-quality Taq polymerase may result in reaction failures. Use only recombinant Taq.
6. Before proceeding to hybridization, check sample amplification by gel electrophoresis.
7. PCR program errors, temperature failures, and poor contact with the PCR block may lead to multiple SSO failures. Perform gel electrophoresis to be sure of sample amplification.
8. For LabType SSO, there is a separate 10-min denaturation/neutralization step prior to hybridization that is applied by the addition of denaturation buffer and neutralization buffers, respectively.
9. For LabType SSO, the hybridization conditions are different. There are three wash steps prior to labeling with SA-PE and one wash step after labeling in order to eliminate unbound probes.
10. A probe mix that is not suspended properly may lead to low bead counts and, as a result, interpretation failures.
11. Poor-quality tubes/pipet tips may lead to the adhesion of the beads to their surface and may cause too-low bead counts.
12. Evaporation during hybridization may lead to random failures in various samples within a batch or batches. In situations in which partial use of plates are planned, keep one row empty on each side of the samples to allow space for tight sealing.
13. Warm DS at 45°C for 5 min upon arrival. Before the labeling, warm the DS to room temperature and vortex, store at room temperature until pipetting. DS/SA-PE mixture must be protected from light.
14. If more than two probes are indeterminate, repeat the assay. If less than two probes are indeterminate, analyze the sample with the probe as negative and repeat with the probe as positive.

    a. One of the choices provide a match. continue to the next step.
    b. Both choices provide matches, reassay the sample.
    c. Neither choices produce a match, check for the other possible incorrectly assigned probes, check for an amplification failure, and repeat the assay.

References

1. Blasczyk, R. (1998) New HLA typing methods, in *New Diagnostic Methods in Oncology and Hematology,* (Huhn, D., ed.), Springer, Berlin, Germany, pp. 143–195.
2. Saiki, R. K., Bugawan,T. L., Horn, G. T., Mullis, K. B., and Erlich, H. A. (1986) Analysis of enzymatically amplified beta globin and HLA-DQ alpha DNA with allele specific oligonucleotide probes. *Nature* **324,** 163–166.
3. Kostyu, D. D., Pfohl, J., Ward, F. E., Lee, J., Murray, A., and Amos, D. B. (1993) Rapid HLA-DR oligotyping by an enzyme-linked immunoabsorbent assay performed in microtiter trays. *Hum. Immunol.* **38,** 148–158.
4. Bugawan, T. L., Apple, R., and Erlich, H. A. (1994) A method for typing polymorphism at the HLA-A Locus using PCR amplification and immobilized oligonucleotide probes. *Tissue Antigens* **44**, 137–147.
5. Middleton, D., Williams, F., Cullen, C., and Mallon, E., (1995) Modification of an HLA-B PCR-SSOP typing system leading to improved allele determination. *Tissue Antigens* **45,** 232–236.
6. Kennedy, L. J., Poulton, K. V., Dyer, P. A., Ollier W. E., and Thomson, W. (1995) Definition of HLA-C alleles using sequence specific oligonucleotide probes (PCR-SSOP). *Tissue Antigens* **46,** 187–195.
7. Cereb, N., Maye, P., Lee, S., Kong, Y., and Yang, S.Y. (1995) Locus specific amplification of HLA Class I genes from genomic DNA: locus specific sequences in the first and third introns of HLA-A, -B and -C, alleles. *Tissue Antigens* **45,** 1–11.
8. Robinson J., Malik., Parham, P., Bodmer, J. G., and Marsh, S. G. E. (2000) IMGT/HLA database a sequence database for the human major histocompatibility complex. *Tissue Antigens* **55,** 280–287.
9. LifeMATCH Website (http://www.lifematchhla.com). Accessed June 2006.
10. http://www.tepnel.com/life_codes/hla_testing_services.asp/
11. One Lambda Inc. (http://www.onelambda.com). Accesed June 2006.
12. Wu, Y. Y. and Csako, G. (2006) Rapid and/or high throughput genotyping for human red blood cell, platelet,and leukocyte antigens, and forensic applications. *Clin. Chimica Acta* **363,** 165–176.
13. Dunbar, A. S. (2006) Applications of Luminex® MAP technology for rapid, high throughput multiplexed nucleic acid detection. *Clinica Chimica Acta* **363,** 71–82.
14. Fulton, R. J., McDade, R. L., Smith, P. L., Kienker, L. J., and Kettman,Jr. J. R. (1997) Advanced multiplexed analysis with FlowMetrix system. *Clin. Chem.* **43,** 1749–1756.

6

Sequencing-Based Typing of HLA

Berthold Hoppe and Abdulgabar Salama

Summary

Human leukocyte antigen (HLA) typing is largely performed by use of PCR-based techniques in patients that require stem cell transplantation. In this chapter, HLA typing by sequencing-based typing techniques is described.

Key Words: HLA typing; SBT; allele-specific sequencing.

1. Introduction

There is no doubt that mismatches, at least for human leukocyte antigen (HLA)-A, -B, -Cw, and -DRB1, have an impact on the outcome of patients that have undergone hematopoietic stem cell transplantation (HSCT), even when solely present on the allelic level ***(1,2)***. Therefore, donor selection should ideally be based on high-resolution typing of HLA classes I and II. This may be achieved by PCR using sequence-specific oligonucleotides (PCR-SSO) and/or sequence-specific primers (PCR-SSP) (*see* Chapters 5 and 4). However, an unambiguous and definite HLA type at the allelic level can only be achieved by sequencing-based typing (SBT) ***(3)***. In the case of HLA class I, it is essential that at least both exons 2 and 3 be sequenced. For HLA-DRB1 typing, exon 2 of DRB1 should be analyzed. On some occasions, sequencing of additional exons is required to achieve correct allele identification. Although the identification of differences between donor and recipient in these additional exons has not yet been shown to have a direct impact on transplantation outcome, nonexpressing alleles due to variations in the corresponding gene could play an important role. Meanwhile, there are a wide variety of SBT kits for high-resolution HLA typing that are commercially available, such as from Atria Genetics (San Francisco, CA) distributed by Abbott Diagnostics (Wiesbaden,

From: *Methods in Molecular Medicine, vol. 134: Bone Marrow and Stem Cell Transplantation*
Edited by: M. Beksac © Humana Press Inc., Totowa, NJ

Germany), Invitrogen (Karlsruhe, Germany), and Protrans (Hockenheim, Germany). Most of these SBT kits allow for high-resolution typing of at least HLA-A, -B, and -DRB1. In addition, several in-house SBT approaches have been published by different groups ***(4–6)***. However, all of these approaches, commercial and noncommercial, differ in their capacity for resolving the *cis-trans* linkage, i.e., in identifying a definite HLA type by separating both alleles. In cases in which both alleles cannot be separated by the amplification or sequencing reaction, the SBT results in a heterozygous DNA sequence. In such cases, only a presumable but not definite high-resolution HLA type can be assigned. With the identification of novel HLA alleles, a recalculation of such heterozygous DNA sequences may be required, and the results have the potential to become unambiguous.

In this chapter, we describe the procedure used in our laboratory examinations for HLA-DRB1 high-resolution typing. The capacity for allele-specific sequencing of this approach is achieved by group-specific amplification (GSA). Typing of other HLA loci is performed in a similar manner. The differences between HLA class I and II typing will be outlined, and the principal differences between GSA-based approaches and those using group-specific sequencing primers (GSSP) will be discussed.

Future advancements will perhaps help elucidate which requirements on the resolution level of HLA typing for HSCT are necessary to achieve the optimal overall benefit.

2. Materials

2.1. SBT Kit

All materials are supplied with the S4 HLA-DRB1 SBT kit (Protrans).

1. Strips with lyophilized group-specific primer mixes (PM). These mixes allow for a resolution of the following DR constellations:

PM1: DRB1*01	PM2: DRB1*15/16
PM3: DRB1*03	PM4: DRB1*04
PM5: DRB1*03/11/13/14	PM6: DRB1*08/11/13/14
PM7: DRB1*03/11/13	PM8: DRB1*12
PM9: DRB1*1301/1302	PM10: DRB1*14
PM11: DRB1*07	PM12: DRB1*08
PM13: DRB1*09	PM14: DRB1*10

2. PCR solution D (PSD).
3. AmpliTaq Gold^{TM} DNA Polymerase 5 U/μL (TaqP)
4. Sequencing primers for exon 2 (DR-E2F; forward), exon 2 (DR-E2R; reverse), exon 2 codon 86 GTG (DR-86V; reverse).
5. Loading buffer (LB).

2.2. Additional Materials

1. High-performance liquid chromatography (HPLC) water.
2. Agarose, molecular biology-grade.
3. Dye terminators (depending on the compatibility with the sequencer used), e.g., Big Dye™ Terminator Cycle Sequencing Kit v1.1/3.1 (BDT, Applied Biosystems), DYEnamic™ ET Terminator Cycle Sequencing Kit (Amersham Biosciences), or CEQ Dye-labeled Terminator Cycle Sequencing Kit (Beckman Coulter).
4. Ethidium bromide stock solution, 10 mg/mL.
5. Exonuclease I/Shrimp Alkaline Phosphatase (ExoSAP-IT™) (Amersham Biosciences) for digestion of amplification reaction products.
6. Sephadex G-50 Fine DNA Grade (Amersham Biosciences) and DyePUR or comparable columns (Protrans) for purification of sequencing reactions.
7. Tris-Borate-EDTA (TBE) buffer.

Comparable materials from other suppliers may also be used.

2.3. Equipment

2.3.1. Pre-PCR Area

1. Multichannel pipettor (eight-channel).
2. Multipettor with combitips (Eppendorf).
3. Pipets and filter tips (0.5–10 μL; 10–100 μL; 100–1000 μL).

2.3.2. Thermal Cycler

Different thermal cyclers are available from multiple suppliers. For example, GeneAmp PCR System 9700 (Applied Biosystems).

2.3.3. Post-PCR Area

1. Automated DNA sequencer and consumables. For example, ABI 310 capillary sequencer (Applied Biosystems).
2. Electrophoresis system for submerged horizontal agarose gel.
3. Multichannel pipettor (eight-channel).
4. Multipettor with combitips (Eppendorf).
5. Pipets and filter tips (0.5–10 μL; 10–100 μL; 100–1000 μL).
6. Power supply.
7. Ultraviolet (UV) spectrophotometer for determination of DNA concentration and purity.
8. UV transilluminator for visualization of the amplification products.
9. Vortexer.

2.3.4. Allele Identification Software

Analysis of the DNA sequences can be performed in several ways. Many manufacturers distribute corresponding software such as uTYPE™ HLA Sequencing software (Invitrogen), Matchmaker™ (Abbott Diagnostics), and Protrans S4 Sequence Pilot™ (Protrans).

3. Methods

3.1. DNA Preparation

Different methods are in use for the extraction of genomic DNA. The optimal method requires that high-quality DNA be provided. DNA concentration should be diluted to 50–100 ng/μL using HPLC water.

3.2. PCR Amplification

A total of 14 different group-specific primer mixes are used for amplification of exon 2 of HLA-DRB1. This allows for the separate sequencing of both HLA-DRB1 alleles in the majority of cases.

1. Master mix (MM): 280 μL PSD + 3 μL TaqP + 20 μL DNA in a 1.5-mL tube.
2. 15 μL MM in each tube of the PCR strip except for negative control.
3. Amplification reaction: thermal cycler profile (final volume: 15 μL):

   Hold: 95°C, 2 min
   15 cycles: 96°C, 40 s/64°C, 1 min/72°C, 2 min
   15 cycles: 96°C, 20 s/60°C, 1 min/72°C, 2 min
   10 cycles: 96°C, 20 s/56°C, 1 min/72°C, 2 min
   Hold: 4°C
4. *Analysis of amplification products:* 5 μL of each amplification reaction + 2 mL LB, electrophoresis (2% agarose gel) and visualisation by UV transillumination.

   The results of analysis by gel electrophoresis provide firsthand information about the quality of the amplification reaction, and the actual HLA-DRB1 type (analogous to HLA typing by amplification refractory mutation system [ARMS]).

3.3. Purification of the Amplification Products

Digestion by ExoSAP-IT is performed to remove residual PCR primers and dNTPs. Compared to column-based purification procedures, this technique is especially useful when preparing a large number of samples. It is ideal that those reaction products of two different positive amplification mixtures, in which either contains one of the alleles separately, be purified and sequenced.

1. Each PCR tube contains: 3 μL ExoSAP-IT + 10 μL amplification product.
2. Incubation in thermal cycler: 37°C, 15 min/80°C, 15 min/4°C

See **Notes 1** and **2**.

3.4. DNA Sequencing

If the separation of both HLA-DRB1 alleles by group-specific amplification reactions is achieved, forward sequencing is sufficient using the primer DR-E2F. Otherwise, additional sequencing reactions using the primers DR-E2R and DR-86V should be performed. The following sequencing reaction mixture is recommended:

1. 2 μL BDT + 6 μL sequencing primer + 2 μL purified amplification product.
2. Thermal cycler profile (final volume: 10 μL):
   Hold: 96°C, 1 min
   25 cycles: 96°C, 10 s/50°C, 5 s/60°C, 4 min
   Hold: 4°C

See **Notes 3** and **4**.

3.5. Purification of Sequencing Reaction Products and Preparation for Capillary Electrophoresis

Standard techniques such as gel filtration using Sephadex G-50 can be used for removing unincorporated dye terminators. For capillary electrophoresis, 2 μL of the purified sequencing reaction is added to 18 μL of HPLC water.

3.6. Capillary Electrophoresis

Analysis of the sequencing reaction can be performed on various types of sequencers. Our laboratory utilizes an ABI 310 Capillary DNA Sequencer.

The following sequencer settings are used for HLA-DRB1:

1. POP-6, 47 cm capillary, mobility file: DT POP6{BD Set-Any Primer}.
2. Run module:
   a. Injection time: 30 s, injection voltage: 2.0 kV.
   b. Collection time: 40 min.
   c. Electrophoresis voltage: 15.0 kV.
   d. Temperature: 50°C.
   e. Syringe pump time: 360 s.
3. Sample sheet settings:
   a. Dye set/primer: DT POP6{BD Set-Any Primer}.
   b. Matrix: Big Dye Matrix (E).
4. Basecaller: ABI-CE1 v3.0.

Examples of the resulting sequencing data of two samples typed as HLA-DRB1*0830,1403 and DRB1*1501,1601, respectively, are represented in **Figs. 1** and **2** *(3)*.

3.7. Sequence Analysis

DNA sequence analysis can be performed by use of various methods. Many manufacturers distribute corresponding software such as uTYPE HLA Sequencing software (Invitrogen), Matchmaker (Abbott Diagnostics), and Protrans S4 Sequence Pilot (Protrans). Our laboratory currently utilises the latter software. This program automatically recognizes the direction of sequencing, and the corresponding locus and exon. Based on the sequence data of both alleles, the correct HLA type is automatically identified in most cases. If there are any inconsistencies in the sequence in comparison to the integrated HLA sequence

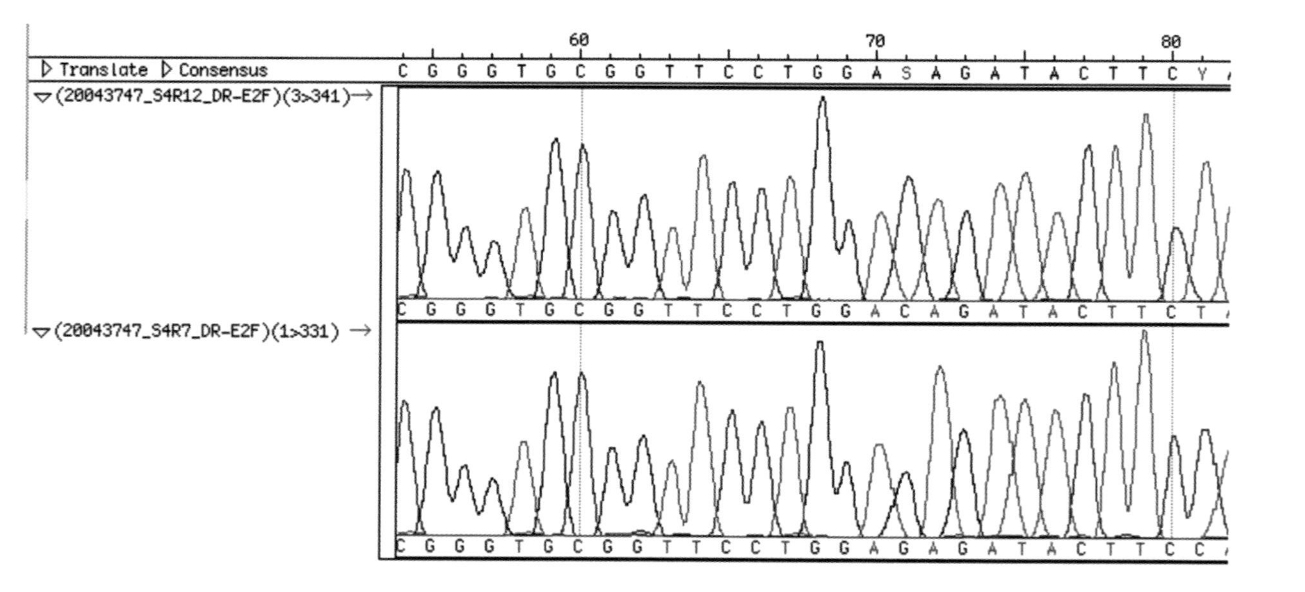

Fig. 1. HLA-DRB1 SBT (DRB1*0830,1403). Both alleles were amplified separately. Two hemizygous sequences of the sequencing reactions with primer DR-E2F representing the alleles DRB1*0830 and DRB1*1403, respectively, are given (SeqMan6.1, DNASTAR Inc.).

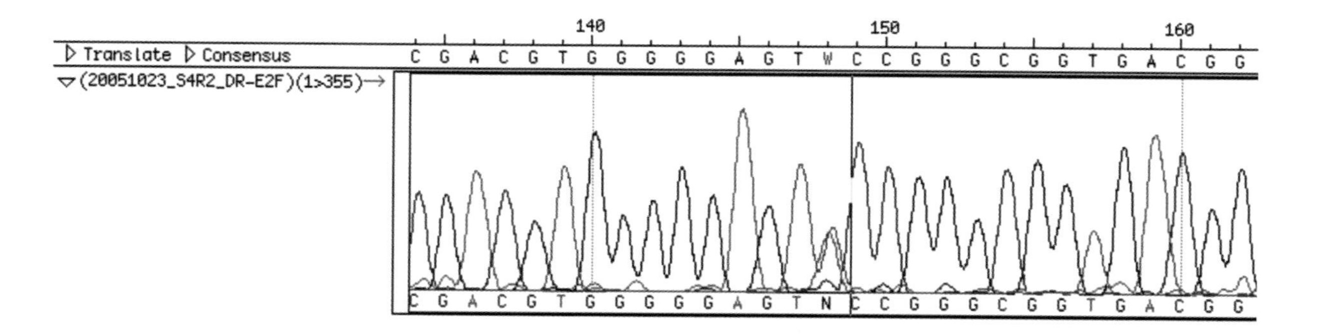

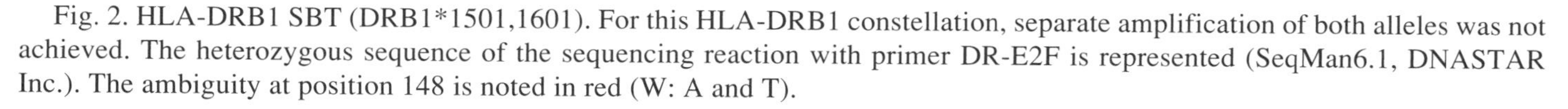
Fig. 2. HLA-DRB1 SBT (DRB1*1501,1601). For this HLA-DRB1 constellation, separate amplification of both alleles was not achieved. The heterozygous sequence of the sequencing reaction with primer DR-E2F is represented (SeqMan6.1, DNASTAR Inc.). The ambiguity at position 148 is noted in red (W: A and T).

database (e.g., owing to unrecognized bases, wide peaks, or novel HLA alleles), the software directs the operator's attention to the discrepant bases for individual evaluation.

4. Notes

1. AlleleSEQR HLA-DRB1 (GSA). When using the AlleleSEQR HLA-DRB1 kit, amplification is performed in a single tube. Where ambiguities are remaining following the sequencing reactions with the sequencing primers exon 2F, exon 2R, and codon 86, a resolution by GSSP (*see* **Note 3**) or GSA should be tried. The GSA-based strategy uses eight different group-specific primer pairs, allowing for specific amplification of DR groups 1, 2, 3/11/6, 4, 7, 8/12, 9, and 10.
2. HLA class I typing by Protrans S4. Depending on the locus to be sequenced, a total of 12 (HLA-A), 14 (HLA-B), or 12 (HLA-Cw) group-specific amplification reactions and 1 single locus-specific amplification reaction should be performed for each locus. The resulting amplification products are considerably larger than those of HLA-DRB1, as both exons 2 and 3 are amplified.
3. AlleleSEQR HLA-DRB1 (GSSP). When using the AlleleSEQR HLA-DRB1 kit (GSSP strategy), ambiguities are primarily resolved by the sequencing reaction. Six different group specific sequencing mixes allow for the performance of specific sequencing reactions for DR groups 1, 2, 3/11/6, 4, 7, and 8/12. Based on the result of the initial AlleleSEQR HLA-DRB1 analysis, adequate sequencing mixes can then be chosen as required.
4. HLA class I typing by Protrans S4. For HLA class I typing, the sequencing primers (X represents A, B, or C) X-E2F, X-E2R, X-E3F, and X-E3R should be used. These primers allow for the complete sequencing of exons 2 and 3. If a PCR-based separation of the alleles is not achieved, the amplification product of the locus-specific amplification reaction should be sequenced using sequencing primers for exons 2 and 3. In some cases, sequencing of exons 1 and 4 (X-E1F, X-E1R, X-E4F, X-E4R) are required.

References

1. Petersdorf, E. W., Anasetti, C., Martin, P. J., et al. (2004) Limits of HLA mismatching in unrelated hematopoietic cell transplantation. *Blood* **104,** 2976–2980.
2. Flomenberg, N., Baxter-Lowe, L. A., Confer, D., et al. (2004) Impact of HLA class I and class II high-resolution matching on outcomes of unrelated donor bone marrow transplantation: HLA-C mismatching is associated with a strong adverse effect on transplantation outcome. *Blood* **104,** 1923–1930.
3. Hoppe, B., Heymann, G. A., Kiesewetter, H., and Salama, A. (2005) Identification and characterization of a novel HLA-DRB1 allele, DRB1*0830. *Tissue Antigens* **66,** 160–162.
4. Swelsen, W. T., Voorter, C. E., and van den Berg-Loonen, E. M. (2005) Sequence-based typing of the HLA-A10/A19 group and confirmation of a pseudogene coamplified with A*3401. *Hum. Immunol.* **66,** 535–542.

5. Kotsch, K., Wehling, J., and Blasczyk, R. (1999) Sequencing of HLA class II genes based on the conserved diversity of the non-coding regions: sequencing based typing of HLA-DRB genes. *Tissue Antigens* **53,** 486–497.
6. van der Vlies, S. A., Voorter, C. E., and van den Berg-Loonen, E. M. (1998) A reliable and efficient high resolution typing method for HLA-C using sequence-based typing. *Tissue Antigens* **52,** 558–568.

7

Molecular Typing Methods for Minor Histocompatibility Antigens

Eric Spierings and Els Goulmy

Summary

Minor histocompatibility (H) antigens crucially affect the outcome of human leukocyte antigen-matched allogeneic stem cell transplantation. The number of molecularly identified minor H antigens is rapidly increasing. In parallel, clinical implementation of minor H antigens for immunotherapy has gained significant interest. It is therefore timely to type stem cell transplant recipients and their donors for minor H antigens. Here, we summarize all the currently known methodologies for minor H antigen typing on the genomic and on the RNA level.

Key Words: Minor histocompatibility antigens; transplantation; stem cells; solid organ; graft-vs-host; graft-vs-tumor; graft rejection; genotyping.

1. Introduction

Minor histocompatibility (H) antigens have been defined as immunogenic peptides derived from polymorphic intracellular proteins and presented on the cell surface by major histocompatibility complex (MHC) class I and class II molecules *(1)*. They induce human leukocyte antigen (HLA)-restricted T-cell responses in the setting of HLA-matched stem cell transplantation (SCT). Minor H antigens must be considered as key molecules in the allo-immune responses of the pathophysiology of graft-vs-host-disease (GvHD) and in the curative graft-vs-tumor (GvT) reactivity after HLA-matched SCT *(2)*. Recently, the role of minor H antigens in renal allograft tolerance was explored. Minor H antigen HA-1-specific regulatory T-cells, HA-1-specific effector T-cells, and HA-1 microchimerism were observed in renal allograft tolerant recipients *(3)*. Interestingly, also in the physiological setting of pregnancy, minor H antigen-specific T-cells could be identified in the mutual direction of mother and child *(4,5)*.

From: *Methods in Molecular Medicine, vol. 134: Bone Marrow and Stem Cell Transplantation*
Edited by: M. Beksac © Humana Press Inc., Totowa, NJ

To date, the clinical usefulness of minor H antigen typing is evident in the area of SCT. As mentioned previously, two clinically related components justify minor H antigen typing, i.e., GvHD and GvT. The mode of tissue expression, i.e., hematopoietic restricted or ubiquitous, of minor H antigens determines their role as target molecules in GvHD and/or GvT. The minor H antigens particularly relevant in the development and maintenance of GvHD are the ones with ubiquitous expression including expression on the GvHD target organs. Minor H antigens with expression limited to cells of the hematopoietic system and to solid tumors are especially relevant for the GvT activity *(6–8)*. The latter minor H antigens can be exploited as curative tools for stem cell-based immunotherapy of hematological malignancies and solid tumors. Hereto, HLA-matched minor H antigen-mismatched recipient/donor combinations are selected. To date, there are two options to exploit the differences in minor H antigen expression between donor and recipients. One strategy is the adoptive transfer of stem cell donor-derived minor H antigen-specific cytotoxic T-lymphocytes (CTLs) generated in vitro. The other, more practical and potentially efficient strategy is post-SCT "vaccination." In this concept, the relevant recipient-specific minor H peptides are administered to boost the donor-derived minor H antigen-specific T-cells to achieve optimal GvT responses in vivo. Both latter strategies are currently ongoing in clinical phase I/II studies *(2)*. Although not frequently observed, minor H antigens have been shown to influence stem cell graft rejection *(9,10)*. In this respect, the expression of minor H antigens on the donor progenitor cells might be relevant in sensitized patients receiving a minor H antigen-disparate T-cell-depleted SCT.

Minor H antigen typing is useful in the clinically related research investigating the role of minor H antigen-specific T-cells in HLA-matched solid organ transplants. The observation of coexistence of CD8+ memory T-effector and -regulator cells, both specific for the same minor H antigen HA-1 in the context of renal allograft tolerance, validates further exploration of the relevance of minor H antigen mismatching in solid organ transplantation *(3)*. The relevance of minor H antigen typing recently further extended toward two types of stem cell donors, i.e., the parous woman and the cord blood donor. Recent examination of multiparous women demonstrated that pregnancy can lead to alloimmune responses against the infant's paternal minor H antigens *(4)*. Interestingly, a similar immunization status has been observed in cord blood samples, where minor H antigen-specific cytotoxic T-cells directed at the noninherited maternal minor H antigens could be demonstrated *(5)*. The minor H antigen immunization status of stem cell donors raises important questions for clinical practice.

Next to minor H antigen geno- and phenotyping, quantification of the minor H antigen gene product at the RNA level may be useful. For some minor H antigens, a quantitative real-time Taqman PCR method for the analysis of minor

Table 1A
Autosomally Encoded Minor H Antigens

Minor H antigen	Peptide	Gene	Distribution	Reference
HA-1	VLHDDLLEA	HA-1	Restricted	*36*
HA-2	YIGEVLVSV	Myosin 1G	Restricted	*20*
HA-3	VTEPGTAQY	LBC Oncogene	Broad	*19*
HA-8	RTLDKVLEV	KIAA0020	Broad	*25*
HB-1	EEKRGSLHVW	Unknown	Restricted	*12*
ACC-1	DYLQYVLQI	BCL2A1	Restricted	*37*
ACC-2	KEFEDDIINW	BCL2A1	Restricted	*37*
UGT2B17	AELLNIPFLY	UGT2B17	Restricted	*27*

Table 1B
Y Chromosome-Encoded Minor H Antigens

Minor H antigen	Peptide	Gene	Distribution	Reference
A1/HY	IVDCLTEMY	USP9Y	Broad	*38,39*
A2/HY	FIDSYICQV	SMCY	Broad	*40*
A33/HY	EVLLRPGLHFR	TMSB4Y	Broad	*41*
B52/HY	TIRYPDPVI	RPS4Y1	Restricted	*42*
B60/HY	RESEEESVSL	UTY	Broad	*43*
B7/HY	SPSVDKARAEL	SMCY	Broad	*44*
B8/HY	LPHNHTDL	UTY	Restricted	*45*
DQ5/HY	HIENFSDIDMGE	DDX3Y DBY	Broad	*46*
DRB1*1501/HY	SKGRYIPPHLR	DDX3Y DBY	Broad	*47*
DRB3*0301/HY	VIKVNDTVQI	RPS4Y1	Broad	*48*

H RNA expression levels has been developed *(11,12)*. Analysis of the membrane expression of minor H antigens by various cell types may be relevant. This can be performed by functional assays such as cell-mediated lympholysis (CML), by growth inhibition of clonogenic normal and leukemic precursor cells (as reviewed in **ref.** ***13***) or by using the skin explant assay *(14)*. **Table 1** lists all currently available information on the human autosomally and Y-chromosome encoded minor H genes, the immunogenic peptides they encode, and the minor H antigen tissue distribution.

In the present chapter we will summarize all methodologies for minor H antigen typing on the genomic and on the RNA level that have been described to date. With regard to the genomic typing, different technologies have been developed by the various laboratories. A universal minor H antigen typing procedure, comprising all minor H antigens, readily available and easy applicable along with the HLA genotyping is currently under construction in our laboratory.

2. Materials

1. 5–10 mL peripheral blood or 5–10 × 10^6 cells.
2. Phosphate-buffered saline (PBS).
3. Ficoll-Isopaque
4. Reagents for DNA isolation.
5. Minor H antigen-specific oligonucleotide primers.
6. Taq polymerase and buffers.
7. PCR temperature cycler.
8. Real-time PCR equipment.
9. Restriction enzymes and buffers.
10. Agarose.
11. Gel electrophoresis equipment.

3. Methods

The methods described next outline (1) the isolation of material for minor H typing; (2) standard PCR procedures; (3) methods for allele-specific PCR-SSP for HA-1, HA-2, and HA-3; (4) PCR using restriction fragment-length polymorphisms (PCR-RFLP) for HA-1, HA-2, HA-8, and HB-1; (5) gene-specific PCR for UGT2B17 and H-Y; (6) real-time PCR for ACC-1 and H-Y; (7) PCR with melting curve analyses for HA-1; and (8) reference strand conformation analysis (RSCA) for HA-1.

3.1. Chromosomal DNA Isolation

Molecular typing for minor H antigens is preferentially performed on genomic DNA above cDNA. For cDNA, RNA must be isolated and cDNA must be synthesized prior to the PCR amplifications. The use of RNA will require extra work and costs. Furthermore, additional care must be taken with the source of the cells used for RNA isolation because the expression of, e.g., the minor H antigens HA-1 and HB-1, can be restricted ***(15,16)***.

All cell types are suitable for genomic DNA extraction. Furthermore, genomic DNA is routinely prepared for HLA typing. Numerous kits and protocols are available for genomic DNA isolation and they all work equally well. One should take into account that it is important that the isolated nucleic acid is free of contaminants and pure enough for PCR purposes. In most cases, genomic DNA will be extracted from peripheral blood derived from peripheral blood mononuclear cells (PBMCs) isolated by Ficoll-Isopaque density gradient centrifugation.

3.2. Polymerase Chain Reaction

Generally, a PCR mix contains a Taq polymerase, a buffer, $MgCl_2$, dNTPs, and primers. Bovine serum albumin can be included to improve performance.

Because concentrations of reagents highly depend on the chosen buffers, enzymes, and methodology, the reader is referred to the original reports and the protocols of the enzyme manufacturer for detailed information. Primers, probes, and restriction enzymes used for minor H antigen typing are listed in separate tables for each methodology.

3.3. Allele-Specific PCR-SSP (see **Note 1**)

PCR–SSP identifies the different alleles by using sequence-specific primers. For optimal performance, the polymorphic nucleic acid is generally located at the last 3′ position of the oligonucleotide primer. This method has been developed for the minor H antigens HA-1 ***(17)***, HA-2 ***(18)***, and HA-3 ***(19)***. The primers for these loci are listed in **Table 2**. PCR amplifications are analyzed by agarose gel electrophoresis.

3.3.1. HA-1 PCR-SSP

HA-1 was the first minor H antigen for which allele-specific PCR-SSP was described ***(17)***. The two allele-specific primers, for HA-1^H and HA-1^R, respectively, are derived from exon A of the HA-1-encoding gene *KIAA0223*. Polymorphic primer sets were designed both in the reverse direction (set 1) and in the forward direction (set 2) in combination with two different common primers (**Table 3**). Cycling condition for the HA-1 SSP-PCR are:

1 cycle	for	5 min	at	95°C
10 cycles	for	1 min	at	95°C
		1 min		65°C
20 cycles	for	1 min	at	95°C
		1 min		62°C
		1 min		72°C

Amplification products of unknown samples are compared with individuals homozygous for either the HA-1^H allele or the HA-1^R allele. The product size for primer set 1 is 190 bp and 331 bp for set 2.

3.3.2. HA-2 PCR-SSP

The HA-2 encoding gene, designated as *MYO1G*, consists of two alleles; HA-2^V and HA-2^M ***(20)***. PCR-SSP has been developed based upon the genomic sequences ***(18)***. The PCR mix for HA-2 contains a common forward primer and polymorphic reverse primers. Cycling condition are:

1 cycle	for	2 min	at	95°C
10 cycles	for	1 min	at	95°C
		1 min		70°C
20 cycles	for	1 min	at	95°C
		1 min		67°C
		1 min		72°C

Both primer combinations result in a PCR product with a size of 274 bp.

Table 2
Oligonucleotide Primers for Allele-Specific PCR Using Sequence-Specific Primers

Minor H antigen	Allele	Forward primer	Reverse primer	Size
HA-1 set 1	H	GTGCTGCCTCCTTGGACACTG	TGGCTCTCACCGTCATGCAG	190
	R	GTGCTGCCTCCTTGGACACTG	TGGCTCTCACCGTCACGCAA	190
HA-1 set 2	H	CTTAAGGAGTGTGTGCTGCA	GCATTCTCTGTTTCCGTGTT	331
	R	CTTAAGGAGTGTGTGTTGCG	GCATTCTCTGTTTCCGTGTT	331
HA-2	V	ACAGTCTCTGAGTGGCTCAG	GCTCCTGGTAGGGGTTCAC	271
	M	ACAGTCTCTGAGTGGCTCAG	GCTCCTGGTAGGGGTTCAT	271
HA-3	T	CTTCAGAGAGACTTGGTCAC	GTTCATGAGCCCATGTTCCAT	129
	M	CTTCAGAGAGACTTGGTCAT	AGACTCAGCAGGTTTGTTAC	318

Table 3
Oligonucleotide Primers and Enzymes for Allele-Specific PCR Using Restriction Fragment-Length Polymorphisms

Minor H antigen	Allele	Forward primer	Reverse primer	Restriction enzyme	Size
HA-1	H	GACGTCGTCGAGGACATCTCCCATC	CTCTTGAGCCAGTGTACGCTCA	Fnu4HI	82+213
				Tsp45I	295
	R	GACGTCGTCGAGGACATCTCCCATC	CTCTTGAGCCAGTGTACGCTCA	Fnu4HI	295
				Tsp45I	85+210
HA-2	V	AAGCTTTTCGAGAAGGGCCGCATCTA	GAATTCGAGATGACGATGCAGGTGTC	*Nla*III or *Hsp*92	218
	M	AAGCTTTTCGAGAAGGGCCGCATCTA	GAATTCGAGATGACGATGCAGGTGTC	*Nla*III or *Hsp*92	163+55
HA-8	R	GGATATACAGCAGAGCTTTC	TCTAACACTTTGTCCCGGAATT	*Eco*RI	165+22
	P	GGATATACAGCAGAGCTTTC	TCTAACACTTTGTCCCAGAATT	*Eco*RI	183
HB-1	H	GAGCCTTCTGACCTCACATC	TTGTCCCTGCTCATCCACACC	*Nla*III	99+135
	Y	GAGCCTTCTGACCTCACATC	TTGTCCCTGCTCATCCACACC	*Nla*III	234

3.3.3. HA-3 PCR-SSP

Genomic typing for HA-3 is based on the sequence of the Rho-GEF *AKAP13* ***(19)***. Polymorphic primers were designed at the same position within the HA-3 gene. They were, however, combined with different reverse oligonucleotides. Cycling condition are:

1 cycle	for	2 min	95°C
10 cycles	for	1 min	95°C
		1 min	70°C
20 cycles	for	1 min	95°C
		1 min	67°C
		1 min	72°C

As a result of the use of different reverse primers, the HA-3^T SSP-PCR yields a product of 129 bp, whereas its negative counterpart HA-3^M has a size of 318 bp. Despite the fact that a single tube analysis is still excluded by this design, misinterpretation of the results is avoided as a result of these differences in product size. Eventually, analysis of the PCR products on agarose gel can be performed by combining the two PCR products into one single lane.

3.4. Allele-Specific PCR-RFLP (see **Note 2**)

The PCR-RFLP identifies point mutations ***(21)*** and polymorphisms ***(22)*** by using restriction endonucleases. In this methodology, the entire polymorphic region of the minor H gene is amplified by PCR first using nonpolymorphic primers. In a second step, an aliquot of the PCR products is digested with the appropriate restriction endonuclease. The products of the restriction digest are analyzed either by agarose gel electrophoresis or by electrophoresis on an acrylamide gel depending on the length of the resulting fragments. PCR-RFLP genomic typing of minor H antigens has been described for HA-1 ***(23)***, HA-2 ***(24)***, HA-8 ***(25)***, and HB-1 ***(12)***.

3.4.1. HA-1 PCR-RFLP

HA-1 PCR-RFLP uses primers to produce a 295-bp product after PCR. Cycling condition for the PCR are:

40 cycles	for	30 s	at	94°C
		30 s		58°C
		1 min		72°C

Products from the HA-1^H allele are susceptible for cleavage by *Fnu4HI*, whereas the HA-1^R product can be digested with *Tsp45I*. Digestions are performed at 37°C for 2 h and analyzed by electrophoresis on 2.2% agarose gels. Successful digestions result in fragments of 85 and 215 bp.

3.4.2. HA-2 PCR-RFLP

For RFLP analysis of HA-2, a fragment with a size of 218 bp is generated by PCR. Cycle parameters are:

denaturation	94°C	2 min	1 cycle
	94°C	1 min	36 cycles
annealing	60.5°C	1 min	
extension	72°C	1 min	
final extension	72°C	10 min	

Samples are digested at 37°C for 12 h with either *Hsp92*II or *Nla*III. The HA-2^V derived sequence will not be digested under these conditions, whereas the HA-2^M allele will result in fragments of 163 and 55 bp.

3.4.3. HA-8 PCR-RFLP

Primers for amplifying the HA-8-encoding gene *KIAA0020* have been designed to result in a 183 bp PCR product. The A at position 20 in the reverse primer has been introduced to artificially create an *Eco*RI site, but only when the primer anneals to the HA-8^R allele. The exact cycling conditions for amplification have not been described. PCR products are digested with *Eco*RI and analyzed on a 2.5% agarose gel. The HA-8^P polymorphism produces a single 183-bp band, whereas the HA-8^R allele will result in bands of 165- and 22-bp.

3.4.4. HB-1 PCR-RFLP

PCR-RFLP for HB-1 is performed on cDNA. The PCR is performed for 33 cycles (1 min at 94°C, 1 min 60°C, and 1 min 72°C). The product has a length of 234 bp. PCR products are digested with *Nla*III to discriminate between HB-1^H and HB-1^Y alleles. The *Nla*III enzyme digests the HB-1^Y allele in two fragments of 99 and 135 bp. The HB-1^H allele will not be cut.

3.5. Gene-Specific PCR

Gene-specific PCR can be applied for the analysis of minor H antigens encoded by genes that lack a negative counterpart. For these minor H antigens, the negative allele cannot be demonstrated directly. Its absence is indicated simply by a negative signal after PCR. Two minor H antigens of this type have been described.

3.5.1. HY PCR

All HY minor H antigens are located on the Y chromosome. Therefore, presence of the Y chromosome implicates positivity for all of the HY minor H antigens. The gene-specific PCR for HY is based on *SRY* ***(26)***. Primers are identical to the ones used in the real-time PCR protocol described under **Subheading 3.6.1.**, except for the fact that no probe is included in the procedure.

Cycling condition are:

1 cycle	for	2 min	at	95°C
10 cycles	for	1 min	at	95°C
		1 min		70°C
20 cycles	for	1 min	at	95°C
		1 min		67°C
		1 min		72°C

Product size for HY-positive samples is 136 bp.

3.5.2. UGT2B17 PCR

Individuals that are negative for UGT2B17 completely lack the *UGT2B17* gene on both chromosomes ***(27)***. As a result of this, this minor H antigen resembles the HY antigens and full genotyping cannot be performed. Three different primer sets have been designed to demonstrate the presence of at least one positive allele; two are located on exon 1, one on exon 6, and one set is designed to amplify the region 5′ to *UGT2B17*. Amplification is performed for 38 cycles genomic DNA or cDNA synthesized from total RNA. Each cycle consists of denaturation (94°C; 30 s), annealing (68°C; 20 s), and elongation (72°C; 30 s). PCR products are analyzed by electrophoresis on a 1.5% agarose gel. The size of the PCR product depends on the selected primer set **(Table 4)**.

3.6. Real-Time PCR (see **Note 3**)

The real-time PCR identifies polymorphisms by the usage of allele specific fluorogenic hybridization probes. The method is based on the 5′ to 3′ exonuclease activity of the Taq DNA polymerase, which results in cleavage of fluorescent dye-labeled probes during PCR ***(28)***. The hybridization probes emit a fluorescent signal that directly correlates with the amount of target sequence. Forward and reverse PCR primers and a probe, labeled with a reporter dye, e.g., FAM or VIC, and a quencher dye, e.g., TAMRA or MGB, bind to the DNA template. A 3′ phosphate group prevents extension of the probe during PCR. The Taq polymerase enzyme enables extension of the primer displacing the probe. The displaced probe is cleaved by Taq DNA polymerase resulting in an increase in relative fluorescence of the reporter dye. The real-time PCR continuously measures the amount of hybridization probe annealing to the target sequence in every cycle.

Rapid genotyping by the real-time PCR method has been described for the detection of the alleles of ACC-1 ***(29)*** and for detection of HY ***(26,30)***.

3.6.1. ACC-1 Real-Time PCR

The fragment covering the ACC-1 polymorphic site of *BCL2A1* can be amplified by primers flanking the polymorphic region. Amplification is performed in the presence of probes complementary to the polymorphic region. These probes are labeled with a fluorescent dye and a quencher as described previously. PCR cycling conditions are: 95°C for 10 min, followed by 35 cycles of 92°C for 15 s and 60°C for 1 min. Amplification and analyses are carried out in a real-time PCR cycler. Because two different fluorochromes are used for each allele, this assay can be performed in a single tube.

3.6.2. HY Real-Time PCR

Y chromosome-specific real-time PCR was developed for measuring the concentration of fetal DNA in maternal plasma or serum *(26)* and for the detection of remaining male cells after sex-mismatched transplantation *(30)*. This methodology can also be applied for sex determination of individuals. Oligonucleotides for amplification are located on the *SRY* gene or the *DFFRY* gene, both situated on the Y chromosome. Amplification primers for *SRY* real-time PCR, which are identical to the gene-specific PCR primers described under **Subheading 3.5.1.**, are combined with a dual-labeled FAM-TAMRA SRY probe containing a 3′-blocking phosphate group (**Table 5**). PCR is initiated with 2 min of 50°C for uracil DNA glycosylase treatment *(31)* (*see* **Note 4**), followed by 10 min of 95°C and 40 cycles of 15 s 95°C and 1 min 60°C.

The described methodology to detect *DFFRY* is similar as for *SRY*. Amplification conditions were as follows: 95°C for 2.5 min, then 30 cycles at 95°C for 45 s, 58°C for 30 s, and 72°C for 1 min, followed by extension at 72°C for 10 min.

3.7. PCR With Melting Curve Analysis

The PCR with melting curve analysis for HA-1 genotyping *(32)* uses fluorescence resonance energy transfer (FRET) occurring between two closely adjacent fluorescent oligonucleotides binding to the same strand of DNA *(33)*. A long oligonucleotide (5′-TTTCTCAAGGCCCTCAGCGAAGCGG-3′) is labeled with fluorescein as donor fluorophore and the shorter one (5′-CTCT CACCGTCATGCAGCACACACTCCTT-3′) is labeled with acceptor fluorophore LightCycler Red 640 (RED640) on the adjacent sites. After binding, the fluorescein is excited by the light source of the LightCycler instrument. A part of this energy is transferred by FRET to RED640, which consecutively emits the measured light of a different wavelength.

Amplification with primers 5′-AGGACATCTCCCATCTGCTG-3′ and 5′-GCATTCTCTGTTTCCGTGTT-3′ is performed using the following conditions: 1 cycle of 30 s at 95°C, followed by 40 cycles of 1 s 95°C, 10 s 64°C, and 20 s 72°C. After amplification of the polymorphic region with flanking primers, melting curve analyses is executed. In this analysis, the PCR product is first denatured at 95°C, followed by a probe-annealing step at 59°C for 15 s.

Table 4
Oligonucleotide Primers for Gene-Specific PCR

Minor H antigen	Forward primer	Reverse primer	Size
UGT2B17			
5′ region	GGCAGTATCTTGCCAATGT	AGACTCCAAGTGCCAGTT	341
exon 1a	TGTTGGGAATATTCTGACTATAA	CCCACTTCTTCAGATCATATGC	352
exon 1b	AAATGACAGAAAGAAACAA	GCATCTTCACAGAGCTTATAT	443
exon 6	GAATTCATCATGATCAACCG	ACAGGCAACATTTTGTGATC	201
HY (SRY)	TGGCGATTAAGTCAAATTCGC	CCCCCTAGTACCCTGACAATGTATT	136

Table 5
Oligonucleotide Primers and Probes for Real-Time PCR

Minor H antigen	Allele	Forward primer	Reverse primer	Probe	Size
ACC-1	Y	ATTTACAGGCTGGCTCAGGA CTA	GGACCTGATCCAGGTTGTGGTAT	CTGCAGTGCGTCCT	65
	C	ATTTACAGGCTGGCTCAGGA CTA	GGACCTGATCCAGGTTGTGGTAT	TCTGCAGTACGTCCTA	65
HY (SRY)	+	TGGCGATTAAGTCAAATTCGC	CCCCCTAGTACCCTGACAATGTATT	AGCAGTAGAGCAGTCA GGGAGGCAGA	136
HY (DFFRY)	+	AACTCACCTCCAACACATACT CCAC	TTCATGATGAAATCTGCTTTTTGTTT	CAGCCACCAGAATTATC TCCAAGCTCTCTGA	84

Subsequently, the temperature is raised to 85°C in steps of 0.1°C per second. Because the RED640-labeled probe has a lower melting temperature, it will be released earlier than the fluorescein-labeled probe. This process can be measured in real-time in FRET analyses. Because the RED640 probe is matched for the HA-1^H allele and has two nucleotides mismatched with the HA-1^R allele, the FRET signal of the HA-1^R allele will show up earlier in the melting process than the HA-1^H allele (67.8 vs 71.5°C, respectively). Although complicated in design, this methodology allows rapid analysis in a single tube.

3.8. Reference Strand Conformation Analysis (see **Note 5**)

The reference strand conformation analysis (RSCA) uses fluorescent-labeled reference DNA (fluorescent-labeled reference [FLR], sense strand) derived from a PCR of the locus of interest *(34)*. This fluorescent reference DNA is hybridized with a locus-specific PCR product from the sample to be tested. Duplexes are formed between the sense and antisense strands present in the mixture. Because only the sense strand is labeled, duplexes formed with this strand can be identified by a laser detection system after electrophoresis in an automated DNA sequencer. Each duplex generated has a unique mobility and therefore each allele is represented. Initially, the RSCA method was used for the detection of new polymorphisms.

The RSCA method has been described as a reliable method for genomic typing of the minor H antigen HA-1 *(35)*. A fragment of 486 bp derived from the HA-1 locus specific region is first amplified by PCR using a common forward primer (5′-GTGCTGCCTCCTTGGACACTG-3′) in combination with a common reverse primer (5′-GCATTCTCTGTTTCCGTGTT-3′). Cycling condition are:

1 cycle	for	5 min	at	95°C
10 cycles	for	1 min	at	95°C
		1 min		65°C
25 cycles	for	1 min	at	95°C
		1 min		62°C
		1 min		72°C

An FLR is generated using HA-1^H and HA-1^R homozygous material using a Cy5-labeled forward primer. Hybridization of the FLR with the PCR product from the samples is performed at a ratio of 1:3. Electrophoresis is executes on 6% nondenaturing polyacrylamide gels in an automated sequencer.

4. Notes

1. A disadvantage of PCR-SSP is the possibility to miss an allele when the amplification conditions are suboptimal. Internal controls should be included in the PCR reaction. In the previously described PCR-SSP protocol for the minor H antigen

HA-1 an internal control has been introduced for the one tube assay. Here, the common reverse and the common forward primer must amplify a band of 486 bp in all samples. Alternatively, primers for a nonpolymorphic gene can be included.

2. PCR-RFLP analysis is a reliable technique. However, it is laborious and time consuming, because multiple manipulations are required for each sample. Furthermore, different endonucleases are required for detecting the alleles of all minor H antigens.
3. The real-time PCR method is a rapid and reliable method without any post-PCR sample manipulation, allowing high-throughput analyses. The method, however, requires the use of expensive equipment. Furthermore, cross-reaction of allele specific oligonucleotide primers can occur and give false-positive signals.
4. Carry-over contamination from previous PCR can be a significant problem as a result of the abundance of PCR products and to the ideal structure of the contaminant material for reamplification. Therefore controls to detect contamination should always be included. An option to control carry-over contamination with PCR products is using uracil DNA glycosylase in all PCR samples *(31)*. The procedure requires the following two steps: (1) incorporating dUTP in all PCR products (by substituting dUPT for dTTP, and (2) treating all subsequent fully preassembled starting reactions with uracil DNA glycosylase (UDG), followed by thermal inactivation of UDG. UDG cleaves the uracil base from uracil-containing DNA, but has no effect on thymidine-containing DNA. This cleavage blocks replication by DNA polymerases. UDG does not react with dUTP, and is inactivated during denaturation at 95°C.
5. With the RSCA method, many samples can be processed at the same time. A major drawback is that the RSCA method requires special equipment and computer programs that are not available in each laboratory.

Acknowledgments

This work was supported by grants from the Leiden University Medical Center, the Dutch Organization for Scientific Research (NWO), the Leukemia and Lymphoma Society of America, and the 6th Framework Programme "Allostem" of the European Commission.

References

1. Goulmy, E. (1996) Human minor histocompatibility antigens. *Curr. Opin. Immunol.* **8,** 75–81.
2. Hambach, L. and Goulmy, E. (2005) Immunotherapy of cancer through targeting of minor histocompatibility antigens. *Curr. Opin. Immunol.* **17,** 202–210.
3. Cai, J., Lee, J., Jankowska-Gan, E., et al. (2004) Minor H antigen HA-1-specific regulator and effector CD8+ T cells, and HA-1 microchimerism, in allograft tolerance. *J. Exp. Med.* **199,** 1017–1023.
4. Verdijk, R. M., Kloosterman, A., Pool, J., et al. (2004) Pregnancy induces minor histocompatibility antigen-specific cytotoxic T cells: implications for stem cell transplantation and immunotherapy. *Blood* **103,** 1961–1964.

5. Mommaas, B., Stegehuis-Kamp, J. A., Van Halteren, A. G., et al. (2004) Cord blood comprises antigen-experienced T cells specific for maternal minor histocompatibility antigen HA-1. *Blood* **105,** 1823–1827.
6. Mutis, T. and Goulmy, E. (2002) Hematopoietic system-specific antigens as targets for cellular immunotherapy of hematological malignancies. *Semin. Hematol.* **39,** 23–31.
7. Klein, C. A., Wilke, M., Pool, J., et al. (2002) The hematopoietic system-specific minor histocompatibility antigen HA-1 shows aberrant expression in epithelial cancer cells. *J. Exp. Med.* **196,** 359–368.
8. Fujii, N., Hiraki, A., Ikeda, K., et al. (2002) Expression of minor histocompatibility antigen, HA-1, in solid tumor cells. *Transplantation* **73,** 1137–1141.
9. Goulmy, E., Termijtelen, A., Bradley, B. A., and van Rood, J. J. (1976) Alloimmunity to human H-Y. *Lancet* **2,** 1206.
10. Voogt, P. J., Fibbe, W. E., Marijt, W. A., et al. (1990) Rejection of bone-marrow graft by recipient-derived cytotoxic T lymphocytes against minor histocompatibility antigens. *Lancet* **335,** 131–134.
11. Wilke, M., Dolstra, H., Maas, F., et al. (2003) Quantification of the HA-1 gene product at the RNA level; relevance for immunotherapy of hematological malignancies. *Hematol. J.* **4,** 315–320.
12. Dolstra, H., Fredrix, H., Maas, F., et al. (1999) A human minor histocompatibility antigen specific for B cell acute lymphoblastic leukemia. *J. Exp. Med.* **189,** 301–308.
13. Goulmy, E. (1997) Human minor histocompatibility antigens: new concepts for marrow transplantation and adoptive immunotherapy. *Immunol. Rev.* **157,** 125–140.
14. Dickinson, A. M., Wang, X. N., Sviland, L., et al. (2002) In situ dissection of the graft-versus-host activities of cytotoxic T cells specific for minor histocompatibility antigens. *Nat. Med.* **8,** 410–414.
15. de Bueger, M. M., Bakker, A., van Rood, J. J., Van der Woude, F., and Goulmy, E. (1992) Tissue distribution of human minor histocompatibility antigens. Ubiquitous versus restricted tissue distribution indicates heterogeneity among human cytotoxic T lymphocyte-defined non-MHC antigens. *J. Immunol.* **149,** 1788–1794.
16. Dolstra, H., Fredrix, H., Preijers, F., et al. (1997) Recognition of a B cell leukemia-associated minor histocompatibility antigen by CTL. *J. Immunol.* **158,** 560–565.
17. Wilke, M., Pool, J., den Haan, J. M., and Goulmy, E. (1998) Genomic identification of the minor histocompatibility antigen HA-1 locus by allele-specific PCR. *Tissue Antigens* **52,** 312–317.
18. Wilke, M., Pool, J., and Goulmy, E. (2002) Allele specific PCR for the minor Histocompatibility antigen HA-2. *Tissue Antigens* **59,** 304–307.
19. Spierings, E., Brickner, A. G., Caldwell, J. A., et al. (2003) The minor histocompatibility antigen HA-3 arises from differential proteasome-mediated cleavage of the lymphoid blast crisis (Lbc) oncoprotein. *Blood* **102,** 621–629.
20. den Haan, J. M., Sherman, N. E., Blokland, E., et al. (1995) Identification of a graft versus host disease-associated human minor histocompatibility antigen. *Science* **268,** 1476–1480.

21. Deng, G. (1988) A sensitive nonradioactive Pcr-Rflp analysis for detecting point mutations at 12Th codon of oncogene C-Ha-Ras in Dnas of gastric-cancer. *Nucleic Acids Res.* **16,** 6231.
22. Mercier, B., Ferec, C., Dufosse, F., and Huart, J. J. (1992) Improvement in Hla-Dqb typing by Pcr-Rflp: introduction of a constant restriction site in one of the primers for digestion control. *Tissue Antigens* **40,** 86–89.
23. Tseng, L. H., Lin, M. T., Martin, P. J., Pei, J., Smith, A. G., and Hansen, J. A. (1998) Definition of the gene encoding the minor histocompatibility antigen HA-1 and typing for HA-1 from genomic DNA. *Tissue Antigens* **52,** 305–311.
24. Pierce, R. A., Field, E. D., Mutis, T., et al. (2001) The HA-2 minor histocompatibility antigen is derived from a diallelic gene encoding a novel human class I myosin protein. *J. Immunol.* **167,** 3223–3230.
25. Brickner, A. G., Warren, E. H., Caldwell, J. A., et al. (2001) The immunogenicity of a new human minor histocompatibility antigen results from differential antigen processing. *J. Exp. Med.* **193,** 195–206.
26. Lo, Y. M. D., Tein, M. S. C., Lau, T. K., et al. (1998) Quantitative analysis of fetal DNA in maternal plasma and serum: implications for noninvasive prenatal diagnosis. *Am. J. Hum. Genet.* **62,** 768–775.
27. Murata, M., Warren, E. H., and Riddell, S. R. (2003) A human minor histocompatibility antigen resulting from differential expression due to a gene deletion. *J. Exp. Med.* **197,** 1279–1289.
28. Heid, C. A., Stevens, J., Livak K. J., and Williams, P. M. (1996) Real time quantitative PCR. *Genome Res.* **6,** 986–994.
29. Akatsuka, Y., Warren, E. H., Gooley, T. A., et al. (2003) Disparity for a newly identified minor histocompatibility antigen, HA-8, correlates with acute graft-versus-host disease after haematopoietic stem cell transplantation from an HLA-identical sibling. *Br. J. Haematol.* **123,** 671–675.
30. Fehse, B., Chukhlovin, A., Kuhlcke, K., et al. (2001) Real-time quantitative Y chromosome-specific PCR (QYCS-PCR) for monitoring hematopoietic chimerism after sex-mismatched allogeneic stem cell transplantation. *J. Hematother. Stem Cell Res.* **10,** 419–425.
31. Longo, M. C., Berninger, M. S., and Hartley, J. L. (1990) Use of uracil DNA glycosylase to control carry-over contamination in polymerase chain reactions. *Gene* **93,** 125–128.
32. Kreiter, S., Wehler, T., Landt, O., Huber, C., Derigs, H. G., and Hess, G. (2000) Rapid identification of minor histocompatibility antigen HA-1 subtypes H and R using fluorescence-labeled oligonucleotides. *Tissue Antigens* **56,** 449–452.
33. Wittwer, C. T., Herrmann, M. G., Moss, A. A., and Rasmussen, R. P. (1997) Continuous fluorescence monitoring of rapid cycle DNA amplification. *Biotechniques* **22,** 130–138.
34. Arguello, J. R., Little, A. M., Pay, A. L., et al. (1998) Mutation detection and typing of polymorphic loci through double-strand conformation analysis. *Nat. Genet.* **18,** 192–194.

35. Arostegui, J. I., Gallardo, D., Rodriguez-Luaces, M., et al. (2000) Genomic typing of minor histocompatibility antigen HA-1 by reference strand mediated conformation analysis (RSCA). *Tissue Antigens* **56,** 69–76.
36. den Haan, J. M., Meadows, L. M., Wang, W., et al. (1998) The minor histocompatibility antigen HA-1: a diallelic gene with a single amino acid polymorphism. *Science* **279,** 1054–1057.
37. Akatsuka, Y., Nishida, T., Kondo, E., et al. (2003) Identification of a polymorphic gene, BCL2A1, encoding two novel hematopoietic lineage-specific minor histocompatibility antigens. *J. Exp. Med.* **197,** 1489–1500.
38. Pierce, R. A., Field, E. D., den Haan, J. M., et al. (1999) Cutting edge: the HLA-A*0101-restricted HY minor histocompatibility antigen originates from DFFRY and contains a cysteinylated cysteine residue as identified by a novel mass spectrometric technique. *J. Immunol.* **163,** 6360–6364.
39. Vogt, M. H., de Paus, R. A., Voogt, P. J., Willemze, R., and Falkenburg, J. H. (2000) DFFRY codes for a new human male-specific minor transplantation antigen involved in bone marrow graft rejection. *Blood* **95,** 1100–1105.
40. Meadows, L., Wang, W., den Haan, J. M., et al. (1997) The HLA-A*0201-restricted H-Y antigen contains a posttranslationally modified cysteine that significantly affects T cell recognition. *Immunity* **6,** 273–281.
41. Torikai, H., Akatsuka, Y., Miyazaki, M., et al. (2004) A novel HLA-A*3303-restricted minor histocompatibility antigen encoded by an unconventional open reading frame of human TMSB4Y gene. *J. Immunol.* **173,** 7046–7054.
42. Ivanov, R., Aarts, T., Hol, S., et al. (2005) Identification of a 40S ribosomal protein S4 - Derived H-Y epitope able to elicit a lymphoblast-specific cytotoxic T lymphocyte response. *Clinical Cancer Research* **11,** 1694–1703.
43. Vogt, M. H., Goulmy, E., Kloosterboer, F. M., et al. (2000) UTY gene codes for an HLA-B60-restricted human male-specific minor histocompatibility antigen involved in stem cell graft rejection: characterization of the critical polymorphic amino acid residues for T- cell recognition. *Blood* **96,** 3126–3132.
44. Wang, W., Meadows, L. R., den Haan, J. M., et al. (1995) Human H-Y: a male-specific histocompatibility antigen derived from the SMCY protein. *Science* **269,** 1588–1590.
45. Warren, E. H., Gavin, M. A., Simpson, E., et al. (2000) The human UTY gene encodes a novel HLA-B8-restricted H-Y antigen. *J. Immunol.* **164,** 2807–2814.
46. Vogt, M. H., van den Muijsenberg, J., Goulmy, E., et al. (2002) The DBY gene codes for an HLA-DQ5 restricted human male specific minor histocompatibility antigen involved in GvHD. *Blood* **99,** 3027–3032.
47. Zorn, E., Miklos, D. B., Floyd, B. H., et al. (2004) Minor histocompatibility antigen DBY elicits a coordinated B and T cell response after allogeneic stem cell transplantation. *J. Exp. Med.* **199,** 1133–1142.
48. Spierings, E., Vermeulen, C., Vogt, M. H., et al. (2003) Identification of HLA class II-restricted H-Y-specific T-helper epitope evoking CD4+ T-helper cells in H-Y-mismatched transplantation. *Lancet* **362,** 610–615.

8

Non-HLA Gene Polymorphisms in Stem Cell Transplantation

Pete G. Middleton

Summary

An increasing number of gene polymorphisms of immune regulatory molecules are being associated with clinical performance following stem cell transplantation (SCT). These polymorphisms affect structural or regulatory changes on immune regulatory molecules including cytokines ***(1–11)***, steroid hormone family receptors ***(12,13)***, pathogen effectors such as mannose binding lectins ***(14)***, and response mediators such as the Fas signaling system ***(15)***.

In contrast to polymorphisms of the major histocompatibility complex, the genome variations found in these non-human leukocyte antigen genes are simple to detect, allowing studies to be done in many laboratories and transplant centres. Many forms of DNA polymorphism detection are now available, allowing even modest laboratories to mount studies of their own. Despite these advances, studies in SCT have a number of problems relating to the complex clinical situation that they study; issues of study design and data interpretation in transplant studies are complex and challenging and are the main limiting factors, which inhibit progress in confirming genetic features which influence the success of SCT.

Key Words: Cytokine gene polymorphism; bone marrow transplantation; non-HLA gene polymorphism.

1. Introduction

A simple Medline search using the term "gene association study" will reveal hundreds of studies a year reporting and discussing genetic effects in nearly every discipline in medicine. Stem cell transplantation (SCT) is no exception, with an increasing number of reports appearing in the literature.

These studies report associations between candidate genes and a wide range of complications seen following SCT including acute graft-vs-host disease

From: *Methods in Molecular Medicine, vol. 134: Bone Marrow and Stem Cell Transplantation*
Edited by: M. Beksac © Humana Press Inc., Totowa, NJ

(aGvHD), chronic graft-vs-host disease (cGvHD), survival, and sensitivity to infections.

This review of available methodology will divide the subject into three headings, under which the relative strengths and weaknesses of the investigative methods, both methodological and experimental, will be discussed; these are:

1. Study design issues, discussed here in the introduction.
2. Gene analysis (technical) issues, under **Subheading 3.**
3. Data analysis issues under **Subheading 4.**

1.1. Study Design Issues

1.1.1. Cohort Studies

The gene association studies published to date on SCT all use a simple cohort-based association study design. Candidate alleles are tested for their association with a clinical outcome, for example, occurrence of GvHD, by dividing a study population into good and bad performers and comparing allele frequency between the two groups *(8)*.

However, a number of features of patients undergoing transplant are already known to affect transplant performance; these include features such as patient age, state of disease (advanced or not) at time of transplant, gender disparity for donor and patient, and degree of HLA disparity, etc. (for plenary reports of the importance of considering these features *see* **refs.** ***16*** and ***17***). In order to control for these confounding variables, two options are available to the study designer. Study groups can be stratified into subgroups of similar risk and analyzed separately (subgroup analysis), or the cohort dataset can include information on these variables and take their effect into account in the final analysis (a multivariant analysis).

The studies reported to date usually present a combination approach to this problem; the study will often restrict itself to a specific subgroup of patients, e.g., human leukocyte antigen (HLA)-identical sibling transplants, but account for the outstanding known variable such as age and GvHD prophylaxis by a multivariant analysis using regression modeling.

The reason for this combined approach is inevitably a pragmatic one because of the limitations of sample size. With so many variables affecting transplant performance, few studies have sufficient power to be viable if the study populations are reduced to groups with homogeneous features.

1.1.2. Case–Control Studies

The alternative type of association study design, the case–control study, has not yet been seen in reported studies on non-HLA gene effects. In this type of study, patients who represent poor performers (the test group) are partnered with good performers who are matched for the confounding variables (age, gender, disease stage, and so on) to provide the control group. Allele frequencies for a candidate gene would be compared for the two groups.

In studies examining the effect of variations in clinical protocol, case–control studies have been used effectively ***(18)***, but this form of study design requires very large patient groups or well organized collaborative groups to recruit sufficient case controls. This requirement effectively excludes most transplant research groups and consequently, case–control studies have not yet been seen in reports of gene associations with transplant outcomes.

1.1.3. Other Study Designs

The cohort study is not the only form of association study design used in gene studies, but the only practical one available to SCT research.

Other genetic studies such as the transmission disequilibrium test ***(19)***, in which risk alleles are tested for prevalence in affected and unaffected offspring (each sibling having an equal chance of inheriting a particular allele from a parent) are not relevant to SCT, as it is extremely rare for two family members to be affected by haemopoietic disease and require treatment by SCT.

Classic family genetic studies of linkage ***(20)*** are also inappropriate for the same reasons: most candidates for therapy with SCT are the sole member of the family affected, so no comparisons can be made.

1.2. Problems With Association Studies

The experience of association studies in other clinical research fields have been mixed. Although many associations have been suggested, many cannot be verified in independent studies.

These frustrations lead to a simple set of criteria being suggested for use in deciding the value of such studies ***(21)***. These criteria suggest that studies should report:

1. Large sample size.
2. Small p values.
3. Report associations that make biological sense and implicate alleles that affect the gene product in a physiologically meaningful way.
4. Contain an initial study and an independent replication.

Although these criteria are clear, they vary in their relevance to studies on SCT: *Large sample size* is relevant to any study, but the complex clinical requirements of SCT mean that most centres transplant relatively few patients on a specific protocol within a reasonable time frame.

Small p *values* also sounds, at first, like a valid criteria; but outcomes (complications) following SCT are known to be multifactorial in cause, with each individual risk factor contributing to the overall risk of a particular complication; i.e., each individual risk factor has a relatively small individual odds ratio, which contributes to an overall larger odds ratio reflecting clinical outcome. In this scenario, p values will usually be relatively large, as the odds ratios used in their calculation are small.

An initial study and an independent replication. Again, this sounds realistic, requiring a verification of the effect reported. In reality, few centers transplant sufficient cases to allow the examination of a comparator group within a reasonable time frame. The result of this has been the appearance of many relatively small, independent studies, which rely for verification on the reports of other groups studying their own cohorts.

One other criterion would be usefully added to this list for any SCT study, that being to report results on *homogeneous study groups*. Reporting biologically meaningful effects requires that the patients in the study population are responding in similar ways. Radically different treatment protocols, such as the use of T-cell depletion or the use of additional immune suppression, would clearly result in quite different biological responses.

The lack of a homogeneous population represents the largest single variable in gene association studies on SCT and the most difficult to mediate against. Most centers transplant patient groups that are heterogeneous for age, gender, disease and stage of disease, availability of a donor, and pretransplant morbidity. In allowing for the severe clinical consequences inherent in treating patients possessed of these features, the transplant teams use protocols that vary with respect to pre-conditioning, stem cell source, prophylaxis for GvHD, and use of post transplant therapies such as donor lymphocyte infusion (DLI; used to therapy disease relapse).

This results in most studies being very heterogeneous with a risk of missing effects that may only occur in a subgroup of the study population, making comparisons between different study populations very difficult.

1.3. Association Studies: Choosing Gene Candidates

All studies reported to date on non-HLA gene polymorphisms in SCT have tested known candidate polymorphisms identified from other studies on immunopathology. Conventional cell biology studies have previously identified a role for many of these molecules in SCT including tumor necrosis factor (TNF)-α, interleukin (IL)-10, and so on. Studies on other immune dysregulatory states, such as Rheumatoid arthritis, systemic lupus erythematosus (SLE), and immune deficiency states, have identified many polymorphisms affecting the production or structure of these molecules.

Research into the effects of these variations on SCT simply brought the two sets of information together to generate gene candidates suitable for testing in SCT.

In selecting gene polymorphisms for study, a few simple criteria are useful:

- Test polymorphisms of genes that have already shown associations with related pathology in similar or related clinical conditions. It is difficult to justify a completely new "fishing trip" on a new target gene if targets showing associations in related situations (e.g., renal transplant) have not been tested.
- Test polymorphisms which display reasonably common allele frequencies to allow useful statistical analysis (this will also be a function of the sample size to be tested).

A number of commercial kits and reagents are available that type for some of the more common single-nucleotide polymorphisms (SNPs) in cytokine genes. For examples, see "One-Lambda" (One Lambda Inc. Canoga Park, CA; www.onelambda.com) and "Pel-Freez" (Pel-Freez Clinical Systems LLC; www.pel-freez.com).

Many of these methods use allele specific techniques (discussed later) and provide typing of alleles previously identified as high or low producers.

However, these relationships between particular alleles and expression are not universally accepted. Considerable numbers of studies continue to present conflicting findings as to the functionality of many of the polymorphisms under study. These conflicts include studies finding completely opposite findings, e.g., for IL-6, in which case conflicting findings are reported for both normal expression measurements ***(22,23)*** and for disease associations ***(1, 10,22,24)***. When the mechanism of action of the polymorphism is generally agreed upon, the mechanism of action within the disease process may be complex, e.g., for the interferon (IFN)-γ A874T polymorphism, the T-allele associates with elevated IFN-γ production ***(25,26)*** and relative resistance from infectious disease ***(27)***, yet associates with reduced GvHD following bone marrow transplant (BMT), probably by acting via a feed back inhibition control circuit ***(1)***.

Consequently, all simplistic "textbook" associations should be treated with caution when planning work of this kind, and in interpreting the results.

1.4. Haplotypes and Alleles

The human genome is composed of regions of strong linkage disequilibrium (LD), including areas of low historical recombination termed "haplotype" blocks. These regions of strong LD have been demonstrated to have low haplotype diversity, with the common haplotypes explaining the majority of common variations across populations.

Furthermore, these haplotypes can be predicted using only a subset of all available (or known) markers. The consequence of this haplotype structure of the genome is that not all available markers at a gene locus need be analyzed to type an individual for genetic variation. A few well chosen markers will reveal the conserved haplotypes involved.

For example, polymorphisms of the vitamin D receptor gene associated with SCT outcome *(12)* form a series of conserved blocks, allowing analysis of the gene with a small number of markers *(28)*.

For access to the public access database of human SNPs, go to http://www.ncbi.nlm.nih.gov/SNP/.

2. Materials

All reagents described are regular commercial reagents and are available from the manufacturer named in the text.

3. Methods

There is a huge range of different technologies available to type DNA polymorphisms. The reasons for choosing a specific method usually reflect an individual laboratory's background, budget, or both.

On technical issues, one of the major concerns to most laboratories is the reproducibility of the technology they use and the reliability of internal controls, especially if the technology makes an absolute, rather than a relative, measurement.

The following examples demonstrate the basic principles involved and discuss the features needed for reliable genotyping in any system, be it manual or automated.

Detection of gene polymorphisms: the SNP. SNPs have been with us for more than 20 yr. Originally detected in Southern blots of DNA cut with restriction enzymes as restriction fragment length polymorphisms (RFLPs), PCR technology has simplified the technique by allowing the target to be amplified and then digested with the restriction enzyme directly. Products are separated on a gel matrix (agarose of larger fragments, polyacrylamide for smaller fragments) to determine whether a restriction site is present.

(For an example, *see* **Fig.1.** and **ref.** ***29***).

3.1. Restriction Enzyme Digestion of PCR Products (see **Notes 1–4**)

1. Prepare PCR products in the usual way.
2. Dispense 5 μL of PCR product into a fresh PCR tube, and add reaction buffer(s), typically 1 μL of 10X buffer supplied with the restriction enzyme plus 3.75 μL of H_2O. Mix gently by tapping the tube.
3. Add 0.25 μL of restriction enzyme, preferably at 10 μ/μL. This will add an excess of enzyme to ensure a limit (nonpartial) digest.
4. Incubate at the temperature recommended by the manufacturer. This is best achieved using the PCR thermal cycler operating on a soak file, as it efficiently incubates the small format tubes used.

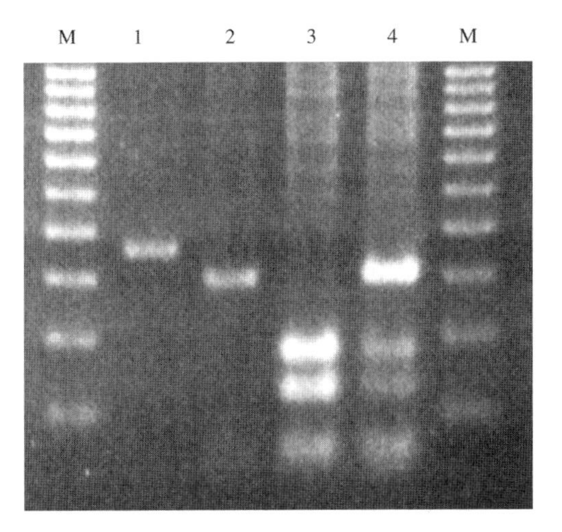

Fig. 1. Typical restriction enzyme digest of PCR products (restriction fragment-length polymorphism [RFLP]). Genotyping of the G-463A polymorphism of the myeloperoxidase gene *(23)*, associate with carcinogen conversion in lung malignancy and infectious complications in a number of pathologies. The 350-bp product possesses an invariant site for the restriction enzyme Aci I (New England Biolabs) plus an additional site in one polymorphic allele. All PCR products must produce a reduced M.wt product on digestion with the additional digestion identifying the G allele possessing the additional restriction site. Lane 1, undigested product; lane 2, AA homozygote; lane 3, GG homozygote; lane 4, AG heterozygote. Molecular weight markers are 100-bp ladder (sigma chemicals), 2% agarose gel (1.5% Microseive agarose, 0.5% SeaKem LE agarose [both Flowgen Instruments Ltd., Sittingbourne, Kent, UK]), staining by ethidium bromide (0.5 µg/mL with visualizatiojn by ultraviolet transillumination). (Photography courtesy of Lindsay Nicholson.)

3.2. Gel Electrophoresis of Restriction Enzyme Digest Products (see **Note 5**)

All but the smallest PCR products can nowadays be analyzed using readily available agarose gels; only very small products require the use of the more complex polyacrylamide gels and equipment.

1. Prepare a 2% (w/v) agarose gel by dispensing 1.5% (w/v) high-resolution agarose (e.g., Microseive Low Melt agarose, FlowGen Co., UK) and 0.5% of any general purpose electrophoresis agarose into a glass conical flask, suspending in the desired buffer volume (the quantities used will depend on the size of the gel electrophoresis equipment). Either of the regularly used molecular biology electrophoresis buffers, TAE or TBE buffer, can be used for this purpose. The general purpose agarose is much stronger mechanically than many of the high resolution agaroses, and makes the resulting gel easier to handle.

2. Weigh the flask and contents and record the weight.
3. Heat the gel mix in a microwave oven until the gel powder/granules begin to dissolve; do this gently to keep the mixture from boiling over, and NEVER look down into the hot liquid.
4. When the gel mixture is homogeneous, return the flask to the balance and replace the weight of lost water with distilled water. Cool the flask to hand warmth and pour the gel into the casting tray. When the gel is cool and set, mount it in the gel electrophoresis tank ready for use.
5. Following electrophoresis of the digests, stain the gel with ethidium bromide to a final concentration of 0.5 μg/mL and visualize the DNA fragments on an ultraviolet (UV) transilluminator, taking precautions as directed by the UV source manufacturer. **Note:** Ethidium bromide is a mutagen; see below for a recommendation for safe disposal.

Pros: Gel electrophoresis is inexpensive, requiring only a thermal cycler and simple gel electrophoresis equipment, and can be carried out in almost any lab.

Cons: Failure or inefficiency of the restriction enzyme step will produce false identification of the undigested allele. Also, the technique is limited to polymorphisms that create or destroy a restriction enzyme recognition sequence. This can be partially augmented by using modifying primers, which create a suitable site in a nearby sequence, but not all polymorphisms are accessible by this technique (for an example used to analyze the TNF-α gene, *see* Wilson et al. *[30]*).

3.3. Allele-Specific Primer PCR (see **Notes 6–8**)

A second method for determining genotype at an SNP is to use allele-specific primers, for which two amplification reactions are run in tandem for each test sample; presence of the allele is indicated by successful amplification, whereas lack of amplification denotes absence of the allele.

The specificity of the PCR to the allelic base difference can be enhanced by manipulating both PCR conditions and the sequence of the allele specific primer pair. This simple experimental design is common to a number of automated and semi-automated genotyping systems as the yes/no output is easily automated; it is, however, sensitive to false-negative results. Failure of a reaction, for example because of poor DNA quality or pipetting error, will result in a sample being incorrectly typed, as the failure of an individual tube to amplify automatically designates the sample a homozygote for the other allele.

In gel-based manual systems, introduction of an additional internal control in the form of a second primer pair to amplify a common (nonpolymorphic) target can reduce this problem (*see* **Fig. 2**).

1. Dispense equal amounts of target DNA into pairs of PCR tubes.
2. Prepare parallel batches of reaction mixes that vary solely by the identity of the allele specific primer.

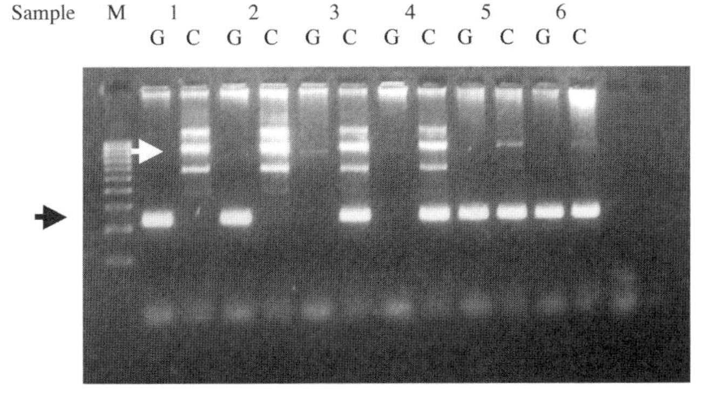

Fig. 2. Allele-specific PCR for the interleukin (IL)-6 G-174C polymorphism *(1)*. G- and C-specific primer reactions run in pairs for each sample. The product of 228 bp (black arrow) denotes the presence of an allele. This reaction produes by chance an internal control in the form of an additional signal in the C reaction (white arrow), which confirms that samples 1 and 2 are GG homozygotes. The lack of suitable internal control in the G tracks makes the call for samples 3 and 4 unreliable (assumes CC homozygotes, but did the G-specific reactions work?). Samples 5 and 6 are heterozygotes (both reactions positive). Marker is a 100-bp ladder (Sigma Chemicals).

3. Dispense mixes (and oil overlay if the thermal cycler requires its use) onto samples.
4. Be careful to arrange reaction tubes in the thermal cycler in matched pairs, so that any variation in performance across the heating block will equally apply to both reactions for a particular target DNA.
5. When completed, analyze reaction products by gel electrophoresis as described above, taking care to run paired reactions adjacent to each other.

   Pros: Allele-specific primer PCR is inexpensive, requiring only a thermal cycler and simple gel electrophoresis equipment, can be carried out in almost any lab, and the simple yes/no format of the output means the technique is easily automated.

   Cons: Failure or inefficiency of one of the pair of reactions will produce false identification of the genotype.

3.4. Single-Stranded Conformation Polymorphism (see **Note 9**)

When a PCR product is heated, the two DNA strands of the helix separate. Upon cooling, some of the duplex will reform, leaving some single stranded DNA molecules unpaired. These single stranded molecules will fold on themselves in order to achieve some stability from base pairing, adopting a particular single-stranded conformation (SSC) as they do so. This conformation is sequence-specific, allowing different alleles to be distinguished from one another *(31)*.

SSCs of DNA have a much more complex structure than the DNA duplex, with mispaired and single-stranded regions inhibiting the molecules' passage through an electrophoresis gel. Consequently, SSCs migrate very slowly compared with the PCR product they are derived from. To allow for resolution of different structures, sufficient migration into a gel matrix must occur; to achieve this, SSC polymorphism assays use polyacrylamide gels with high ratios of acrylamide to bis-acrylamide, typically 37.5:1 (e.g., ProtoFLOWgel Flowgen, Findel House, Ashby de la Zouch, Leicestershire, LE65 1NG, UK). The concentration of gel needed is determined empirically, with gel concentrations between 8 and 16% being the range needed by most targets (*see* **Fig. 3**).

3.5. General Protocol for Preparing Polyacrylamide Gels (see **Note 10**)

The values given below describe preparation of a typical gel mixture (100 mL) suitable for most electrophoresis equipment, e.g., Bio-Rad protean II, Dcode, or Sequi-Gen/Sequi-Gen GT DNA sequencing equipment, and should be scaled to conserve reagents. Best results are achieved using gel spacers and loading well formers of 0.7 mm separation.

1. Dispense the desired amount of acrylamide solution into a 100-mL plastic measuring cyclinder; typically, this will be for a 12% final gel concentration. **Note:** *see* manufacturers information on the properties of acrylamides; they are neurotoxins and potential carcinogens.
2. Add 10 mL of 10X TBE buffer (108 g Tris base, 55 g boric acid, 9.3 g Na_2EDTA, water to 1 L).
3. Add water to 100 mL, dispense into a glass beaker, and de-gas on a vacuum pump (water pump or mechanical oil pump) for 10 min (oxygen inhibits the polymerization).
4. Prepare a fresh 50% (w/v) solution of ammonium persulphate (BDH, Poole, UK) in water (use a 1.5-mL microfuge tube as a weighing boat and prepare about 0.5 mL).
5. To initiate the polymerization, add 1 µL/mL of *NNN-N′*-tetramethylethlenediamine (TEMED; BDH, Poole,England) and 2 µL/mL of the ammonium persulphate solution to the gel mix, fill a large syringe with the initiated mixture, and dispense between the electrophoresis plates. The gel will visibly polymerize within 5 min, but should be allowed to continue polymerising for another 30 min before use.

3.6. General SSC Polymorphism Protocol (see **Notes 11** and **12**)

1. Mix equal volumes of a PCR product with formamide gel loading buffer (9 mL formamide, 1 mL gel loading buffer (50% glycerol, 100 m*M* EDTA), 0.1% bromophenol blue in a thermal cycler tube.

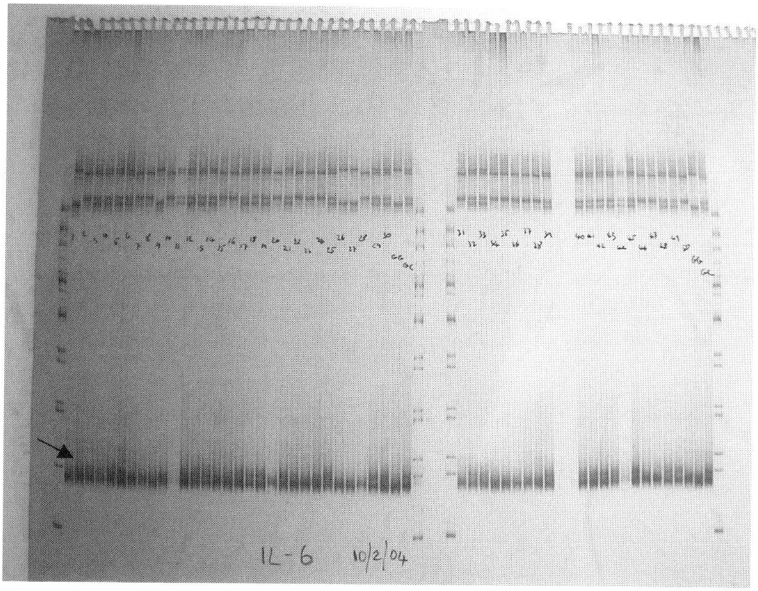

Fig. 3. Single-stranded conformation polymorphism assay for the interleukin (IL)-6 G-174C polymorphism. Using a pair of primers that flank the region of interest (forward primer 5′-TGTCAAGACATGCCAAAGTGCTG-3′, reverse primer 5′ATA GAGCTTCTCTTTCGTTCCCG-3′, annealing temperature 57°C). The product of 245 bp produces unambiguous identification of the alleles by SSCP assay in a 10% polyacrylamide gel (37.5:1 = acrylamide:bis-acrylamide) run overnight at room temperature (J. Norden, personal communication). Sample 1, CC homozygote; sample 2, GG homozygote; sample 3, CG heterozygote. Re-annealed PCR DNA duplexes are seen as conventional PCR product at the bottom of the gel. The additional signal seen above the native duplex PCR product (black arrow) is the heteroduplex caused by mispairing of strands from the two alleles following the melt-anneal step and consequently is seen only in heterozygotes. Gel staining by silver staining.

2. Heat to 95°C for 2 min in a thermal cycler.
3. Remove tubes and plunge directly into ice water.
4. Load onto desired gel. Large sequencing format gels can be run overnight (400 V), whereas smaller formats such as protean equipment can be run over the day using cooling.

   Pros: The SSC polymorphism protocol can be run cheaply, can be applied to most, but not all, SNPs, and can detect linked haplotypes for SNPs present on the same PCR product. Different haplotypes can produce different mobility SSCs, allowing direct determination of haplotype in a single assay. Also, an internally controlled PCR product is always present in the final analysis with its position in the gel determining the allele ID.

Cons: The technique requires control of the gel temperature to exploit its full potential.

3.7. Silver Staining for DNA in Polyacrylamide Gels

Note: This is a "kitchen sink" style method reminiscent of home photography. High-quality reagents are not required.

1. Make up the following solutions immediately before use:

   Solution A (fixer): 100 mL ethanol, 5 mL acetic acid; make up to 1 L with dH_2O.

   Solution B (silver stain): 0.5 g silver nitrate ($AgNO_3$). Make up to 500 mL with dH_2O.

   Solution C (developer): 6 g NaOH (xodium hydroxide), make up to 500 mL with dH_2O; add 0.6 mL formaldehyde solution (Formalin); mix well until NaOH has dissolved.
2. Dismantle gel rig and place plates on the bench, face down.
3. Separate plates, making sure that the gel sticks to the bottom plate.
4. Pour half of solution A onto the gel to allow the gel to float into the tray.
5. Leave this for 3 min, with gentle agitation.
6. Remove fixing solution.
7. Add the second 500 mL of solution A for a further 3 min.
8. Remove as before.
9. Add solution B into the gel tray, agitate for 15 min. Ensure that the solution covers all the gel.
10. Remove as much of the silver nitrate solution as possible, **do not rinse**.
11. Add solution C and agitate for approx 20 min. The DNA signals will become visible during this time.
12. Remove solution C before adding tap water to rinse.

The gel can then be sealed in plastic or dried for storage.

3.8. Disposal of Ethidium Bromide-Containing Buffers

Ethidium bromide is a mutagen and should not be released in to the environment. Safe recommended disposal is by high-temperature incineration. The following method allows safe disposal of ethidium bromide present in aqueous buffers, e.g., electrophoresis buffer.

1. Obtain a large (10 L+) plastic bottle, e.g., "carbuoy" bottle, and place a large plastic funnel in the neck.
2. Dispense approx 50 g of granular charcoal into the bottle.
3. When disposing of contaminated buffers, pour the buffers into the bottle.
4. When the bottle is three-fourths full, mix the contents by gentle swirling for 2–5 min—this binds the ethidium bromide to the charcoal (this can be demonstrated by the loss of an ethidium bromide signature from the buffer, as measured on UV spectrophotometer—just set the photometer to scan a contaminated buffer in the

range 240 to 500 nm).

5. Dispose of the decontaminated buffers via conventional aqueous waste route.
6. Package used charcoal into the solid waste route used for incinerator waste—the ethidium bromide now adsorbed on the charcoal is incinerated as solid waste.

3.9. Data Analysis Issues

A number of separate issues contribute to difficulties in assessing the gene polymorphism dataset once it is collected. Nearly all concern the complexity of the clinical data rather than the gene polymorphism data.

3.9.1. Multiple Features Contributing to Outcomes

Once obtained, gene and clinical datasets are compared to identify genetic features associating with clinical performance or outcome. This simple task is usually achieved by comparing data in a contingency table with statistical assessment by chi-square or similar test modified for small sample size. Almost any statistical package on a PC will perform this analysis.

Considering the effect of other variables known to contribute to the outcome is much more difficult. The size of most studies (<200 transplants) precludes the division of the group into subgroups homogeneous for age, disease stage, and so on; instead, the dataset is used to construct mathematical models of prediction, usually by regression modeling. Widely available statistics packages which easily perform these manipulations include programs such as GraphPad "Prism"(GraphPad Software at www.GraphPad.com), and SPSS (SPSS Inc. 233S Wacker Drive, Chicago, IL).

Pros: This analysis is simple to do on a PC.

Cons: These statistics packages are good for a few features, but many datasets are still too small to allow modeling with the large number of variables present in many SCT studies (*see* study design issues, discussed in the **Subheading 1.**).

When attempting to begin these sorts of analysis consider the following:

1. For most studies, reducing each variable to a "yes" or "no" (binomial regression modeling) reduces the potential number of combinations available to the mathematical model, and is the usual method of choice for simple studies.
2. When assigning binomials (yes or no) to variables prior to analysis be clear as to what your definition of yes and no are; these must match your clinical/biological criteria set out in the study introduction/design.
3. As most SCT cohorts are heterogeneous for known important clinical features, use of multivariant analysis can identify new features by "correcting" the dataset for confounding variables—e.g., the IL-6 G^{-}174C polymorphism represents one of the more reproducible gene markers associating with GvHD, yet both the initial studies that identified this relationship found only trends toward association

using univariant analysis, with the full statistical significance only being revealed following multivariant analysis to "correct" for the noise of the other factors ***(1,10)***.

3.9.2. Stratification and Confounding

Stratification describes where subgroups exist within a dataset. However, the statistical models used to analyze the dataset assume random assortment of the variables; consequently, the statistical significance assigned to a feature or variable in a study may not be valid.

Considerable stratification exist in most SCT study populations. Clinical variables such as preconditioning, modality of therapy, and GvHD prophylaxis are not assigned randomly but are often used on particular patient groups including patients identified by age and disease stage (i.e., they are "elected," not randomized as in a randomized, controlled trial). This lack of separation of variables (confounding) means that even within a relatively large dataset, the contribution of specific features to an overall outcome will often be difficult to measure. Even an independent study may not resolve this sort of problem; patients are treated on the basis of perceived clinical need, and that need, for example the need for increased GvHD prophylaxis in older patients, is likely to be just as evident in different transplant centers.

3.9.3. Censoring and Time-Dependent Measurements

As patients are entered into transplant studies continually, they have all been resident on the study for different periods of time. For outcomes that are time-related, such as death or chronic GvHD, this difference must be accounted for in the analysis—effectively by censoring. This is the conventional form of analysis for survival (death yes or no) but it is increasingly being used for other applications.

This issue is exemplified if we consider looking for gene associations with chronic GvHD. cGvHD can occur up to 2 yr posttransplant. Patients will continue to be deleted from the cohort over this time as a result of death (all causes), without necessarily developing GvHD symptoms during that time. This lack of censoring of the dataset can be accounted for by a change in statistical methodology. The simplest approach, and the one most widely available to a researcher running their own statistical analysis in their group, is to assess cGvHD as a time-dependent, not categorical, variable. When comparing genotypes, death (all causes) should then be included as a competing covariable for the dependent (cGvHD) outcome. Relatively few statistical packages allow for this level of sophistication. A simple alternative used in some studies is to censor patients at point of death in the time dependent analysis. They cease to contribute to the study, and the statistics, from that point on.

It should be noted that relatively few statistics packages allow for multiple regression analysis for a time-dependent outcome such as Cox regression analysis, e.g., a number of features all contributing to death. This is a serious handicap for anyone doing studies on SCT and should direct the experimenter to choosing the statistics package that provides this facility, e.g., the SPSS package.

4. Notes

1. Restriction enzyme digestion conditions: most restriction enzymes are sensitive to reaction conditions; not all will be compatible with the buffer carried over from the PCR reaction. In these cases, precipitate the PCR product with ethanol and resuspend the product in the new buffer.
2. Restriction enzymes are supplied in a buffer containing 50% glycerol; ensure that the final concentration of glycerol in the digest reaction does not exceed 5%, as many enzymes change their characteristics. If more is needed, increase the total reaction volume with buffer and water.
3. When digesting small volumes, consider adding a mineral oil overlay to the reaction just as one would do with PCR reactions, as evaporation with concentrate the reaction salt conditions and alter enzyme properties.
4. Many restriction sites can be recognized by more than one enzyme (isoschizomers). Check the availability of other enzymes, especially any with additional sites (invariant, nonpolymorphic sites) in the PCR product, which will act as internal technical controls to confirm digestion (*see* **Fig. 1**).
5. To speed up the process, ethidium bromide stain can be added to the gel at the point of casting the gel; however the ethidium ions migrate towards the negative electrode and thus leave the gel. To supplement this, ethidium bromide must be added to the portion of the electrophoresis chamber holding the negative electrode to prevent a "tide mark" in the gel where the stain is absent.
6. When designing an allele-specific reaction, remember that either strand of the helix can be used to nominate the specific primer site (i.e., the allele specific primers do not have to be identical to the sense strand!).
7. New assays using allele-specific design must be set up and verified with test samples of known genotype—this is not an investigative tool.
8. Be disciplined in the criteria used for interpreting the final results, especially where controls are involved (e.g., *see* **Fig. 2**).
9. Many PCRs reported to be useful for restriction enzyme digestion to reveal an RFLP will also be suitable for a SSC polymorphism detection; the upper limit of such detection is about 450 bp. PCR products larger than this rarely resolve in the high concentration polyacrylamide matrix.
10. Polyacrylamide gels are very stable once prepared, provided they are always covered in buffer to prevent desiccation, which occurs rapidly for any gel matrix (commercial gels can be purchased precast, ready for mounting directly into the electrophoresis equipment).

11. A number of additional changes can be made to the gel conditions to improve the resolution of some conformers, specifically adding glycerol to the gel (anything to a final concentration of 10% w/v) ***(32)***. Simply add the glycerol (AnalaR grade) in place of the equivalent volume of water to the gel mix prior to adding the TEMED and ammonium persulphate to initiate polymerisation.
12. Stable, sensitive staining of polyacrylamide gels is easily done using simple silver staining. Unlike other forms of silver staining for biomolecules, the DNA does not couple or bond the silver ion, but associates by simple charge attraction (Ag^+ DNA^-). The position of the associated silver is revealed by a simple precipitation of metallic silver by a reaction analogous to a classical photographic developer.

Acknowledgments

Thanks to Jean Norden for the details of the IL-6 SSCP and to Lindsay Nicholson for the details of the Myeloperoxidase RFLP and photograph.

References

1. Cavet, J., Dickinson, A. M., Norden, J., Taylor, P. R., Jackson, G. H., and Middleton, P. G. (2001) Interferon-gamma and interleukin-6 gene polymorphisms associate with graft-versus-host disease in HLA-matched sibling bone marrow transplantation. *Blood* **98,** 1594–1600.
2. Cavet, J., Middleton, P. G., Segall, M., Noreen, H., Davies, S. M., and Dickinson, A. M. (1999) Recipient tumor necrosis factor-alpha and interleukin-10 gene polymorphisms associate with early mortality and acute graft-versus-host disease severity in HLA-matched sibling bone marrow transplants. *Blood* **94,** 3941–3946.
3. Cullup, H., Dickinson, A. M., Cavet, J., Jackson, G. H., and Middleton, P. G. (2003) Polymorphisms of interleukin-1alpha constitute independent risk factors for chronic graft-versus-host disease after allogeneic bone marrow transplantation. *Br. J. Haematol.* **122,** 778–787.
4. Cullup, H., Dickinson, A. M., Jackson, G. H., Taylor, P. R., Cavet, J., and Middleton, P. G. (2001) Donor interleukin 1 receptor antagonist genotype associated with acute graft-versus-host disease in human leucocyte antigen-matched sibling allogeneic transplants. *Br. J. Haematol.* **113,** 807–813.
5. Ishikawa, Y., Kashiwase, K., Akaza, T., et al. (2002) Polymorphisms in TNFA and TNFR2 affect outcome of unrelated bone marrow transplantation. *Bone Marrow Transplant* **29,** 569–575.
6. Keen, L. J., DeFor, T. E., Bidwell, J. L., Davies, S. M., Bradley, B. A., and Hows, J. M. (2004) Interleukin-10 and tumor necrosis factor alpha region haplotypes predict transplant mortality after unrelated donor stem cell transplantation. *Blood* **103,** 3599–3602.
7. MacMillan, M. L., Radloff, G. A., DeFor, T. E., Weisdorf, D. J., and Davies, S. M. (2003) Interleukin-1 genotype and outcome of unrelated donor bone marrow transplantation. *Br. J. Haematol.* **121,** 597–604.

8. Middleton, P. G., Taylor, P. R., Jackson, G., Proctor, S. J., and Dickinson, A. M. (1998) Cytokine gene polymorphisms associating with severe acute graft-versus-host disease in HLA-identical sibling transplants. *Blood* **92,** 3943–3948.
9. Rocha, V., Franco, R. F., Porcher, R., et al. (2002) Host defense and inflammatory gene polymorphisms are associated with outcomes after HLA-identical sibling bone marrow transplantation. *Blood* **100,** 3908–3918.
10. Socie, G., Loiseau, P., Tamouza, R., et al. (2001) Both genetic and clinical factors predict the development of graft-versus-host disease after allogeneic hematopoietic stem cell transplantation. *Transplantation* **72,** 699–706.
11. Stark, G. L., Dickinson, A. M., Jackson, G. H., Taylor, P. R., Proctor, S. J., and Middleton, P. G. (2003) Tumour necrosis factor receptor type II 196M/R genotype correlates with circulating soluble receptor levels in normal subjects and with graft-versus-host disease after sibling allogeneic bone marrow transplantation. *Transplantation* **76,** 1742–1749.
12. Middleton, P. G., Cullup, H., Dickinson, A. M., et al. (2002) Vitamin D receptor gene polymorphism associates with graft-versus-host disease and survival in HLA-matched sibling allogeneic bone marrow transplantation. *Bone Marrow Transplant* **30,** 223–228.
13. Middleton, P. G., Norden, J., Cullup, H., et al. (2003) Oestrogen receptor alpha gene polymorphism associates with occurrence of graft-versus-host disease and reduced survival in HLA-matched sib-allo BMT. *Bone Marrow Transplant* **32,** 41–47.
14. Mullighan, C. G., Heatley, S., Doherty, K., et al. (2002) Mannose-binding lectin gene polymorphisms are associated with major infection following allogeneic hemopoietic stem cell transplantation. *Blood* **99,** 3524–3529.
15. Mullighan, C., Heatley, S., Doherty, K., et al. (2004) Non-HLA immunogenetic polymorphisms and the risk of complications after allogeneic hemopoietic stem-cell transplantation. *Transplantation* **77,** 587–596.
16. Doney, K., Hagglund, H., Leisenring, W., Chauncey, T., Appelbaum, F. R., and Storb, R. (2003) Predictive factors for outcome of allogeneic hematopoietic cell transplantation for adult acute lymphoblastic leukaemia. *Biol. Blood Marrow Transplant* **9,** 472–481.
17. Gratwohl, A., Hermans, J., Goldman, J. M., et al. (1998) Risk assessment for patients with chronic myeloid leukaemia before allogeneic blood or marrow transplantation. Chronic Leukemia Working Party of the European Group for Blood and Marrow Transplantation. *Lancet* **352,** 1087–1092.
18. Diaconescu, R., Flowers, C. R., Storer, B., et al. (2004) Morbidity and mortality with nonmyeloablative compared with myeloablative conditioning before hematopoietic cell transplantation from HLA-matched related donors. *Blood* **104,** 1550–1558.
19. Spielman, R. S., McGinnis, R. E., and Ewens, W. J. (1993) Transmission test for linkage disequilibrium: the insulin gene region and insulin-dependent diabetes mellitus (IDDM). *Am. J. Hum. Genet.* **52,** 506–516.
20. Ghosh, S. and Collins, F. S. (1996) The geneticist's approach to complex disease. *Ann. Rev. Med.* **47,** 333–353.

21. (1999) Freely associating. *Nat. Genet.* **22,** 1–2.
22. Brull, D. J., Montgomery, H. E., Sanders, J., et al. (2001) Interleukin-6 gene-174g>c and -572g>c promoter polymorphisms are strong predictors of plasma interleukin-6 levels after coronary artery bypass surgery. *Arterioscler. Thromb. Vasc. Biol.* **21,** 1458–1463.
23. Fishman, D., Faulds, G., Jeffery, R., et al. (1998) The effect of novel polymorphisms in the interleukin-6 (IL-6) gene on IL-6 transcription and plasma IL-6 levels, and an association with systemic-onset juvenile chronic arthritis. *J. Clin. Invest.* **102,** 1369–1376.
24. Jones, K. G., Brull, D. J., Brown, L. C., et al. (2001) Interleukin-6 (IL-6) and the prognosis of abdominal aortic aneurysms. *Circulation* **103,** 2260–2265.
25. Pravica, V., Asderakis, A., Perrey, C., Hajeer, A., Sinnott, P. J., and Hutchinson, I. V. (1999) In vitro production of IFN-gamma correlates with CA repeat polymorphism in the human IFN-gamma gene. *Eur. J. Immunogenet* **26,** 1–3.
26. Pravica, V., Perrey, C., Stevens, A., Lee, J. H., and Hutchinson, I. V. (2000) A single nucleotide polymorphism in the first intron of the human IFN-gamma gene: absolute correlation with a polymorphic CA microsatellite marker of high IFN-gamma production. *Hum. Immunol.* **61,** 863–866.
27. Rossouw, M., Nel, H. J., Cooke, G. S., van Helden, P. D., and Hoal, E. G. (2003) Association between tuberculosis and a polymorphic NFkappaB binding site in the interferon gamma gene. *Lancet* **361,** 1871–1872.
28. Nejentsev, S., Godfrey, L., Snook, H., et al. (2004) Comparative high-resolution analysis of linkage disequilibrium and tag single nucleotide polymorphisms between populations in the vitamin D receptor gene. *Hum. Mol. Genet.* **13,** 1633–1639.
29. London, S. J., Lehman, T. A., and Taylor, J. A. (1997) Myeloperoxidase genetic polymorphism and lung cancer risk. *Cancer Res.* **57,** 5001–5003.
30. Wilson, A. G., di Giovine, F. S., Blakemore, A. I., and Duff, G. W. (1992) Single base polymorphism in the human tumour necrosis factor alpha (TNF alpha) gene detectable by NcoI restriction of PCR product. *Hum. Mol. Genet.* **1,** 353.
31. Orita, M., Iwahana, H., Kanazawa, H., Hayashi, K., and Sekiya, T. (1989) Detection of polymorphisms of human DNA by gel electrophoresis as single-strand conformation polymorphisms. *Proc. Natl. Acad. Sci. USA* **86,** 2766–2770.
32. Glavac, D. and Dean, M. (1993) Optimization of the single-strand conformation polymorphism (SSCP) technique for detection of point mutations. *Hum. Mutat.* **2,** 404–414.

9

Polymorphisms Within Epithelial Receptors

NOD2/CARD15

Julia Brenmoehl, Ernst Holler, and Gerhard Rogler

Summary

Genetic risk assessment in the setting of allogeneic stem cell transplantation is one of the major goals to optimise future prophylaxis and treatment of patients: our group has focused on analysis of single-nucleotide polymorphisms (SNPs) within the intracytoplasmatic receptor NOD2/CARD15, which recognizes the bacterial cell wall compound muramyl-dipeptide and induces nuclear factor-κB-mediated inflammation. By performing TaqMan PCR of the three major SNPs also identified as risk factors in Crohn's disease in donors and recipients, we were able to demonstrate a major association of NOD2/CARD15 SNPs with the occurrence of severe graft-vs-host disease and resulting treatment-related mortality following human leukocyte antigen-identical sibling transplantation. Although these data need confirmation in further prospective trials, this association may not only be used for risk assessment but also point to a major pathophysiological interaction of dysregulated activation of the innate immune system and specific alloreaction.

Key Words: Innate immunity; graft-vs-host disease; GvHD; NOD2/CARD15.

1. Introduction

Graft-vs-host disease (GvHD) and associated complications are the major cause of morbidity and mortality following allogeneic transplantation, and they occur in spite of optimal human leukocyte antigen (HLA)-matching and immunosuppressive prophylaxis. This clearly indicates the presence of further non-HLA genetic determinants of GvHD, and as discussed elsewhere in this book, cytokine gene polymorphisms have been the first single-nucleotide polymorphisms (SNPs) analyzed and described as important risk factors in the pathogenesis of GvHD *(1)*. In the actual concept of GvHD *(2)*, activation of antigen presenting cells and host epithelia via danger signals delivered either by condi-

From: *Methods in Molecular Medicine, vol. 134: Bone Marrow and Stem Cell Transplantation*
Edited by: M. Beksac © Humana Press Inc., Totowa, NJ

tioning or by microbia is an important step. Only recently, an intracytoplasmatic counterpart to the Toll-like membrane-bound receptors of endotoxin and related bacterial cell wall compounds has been described, the *NOD2/CARD15* molecule. *NOD2/CARD15* belongs to a rapidly growing family of intracytoplasmatic molecules detecting microbial patterns. In 2001, SNPs within the leucine-rich repeat (LRR) region of this molecule, which binds to bacterial cell wall compounds, have been described as the first genetic risk factors of Crohn's disease (CD) ***(3,4)***. As a disease characterized by an uncontrolled Th1 mediated inflammation, CD has pathophysiological overlap with gastrointestinal GvHD. Based on this overlap and the fact that intestinal microbia have been implicated in the pathophysiologiy of GvHD for a very long time, we have started to analyze the relevance of *NOD2/CARD15* SNPs in the recipient or donor in the setting of allogeneic stem cell transplantation. PCR typing for the three major SNPs—8,12, and 13—of *NOD2/CARD15* as described for CD ***(5)*** was correlated with clinical outcome, and we were able to demonstrate a highly significant association of mutated SNPs, although heterozygous in almost all tested patients, with severe gastrointestinal and overall GvHD, but also with 1-yr transplant-related mortality (TRM) ***(6)***. In a recently finished confirmatory study including more than 300 HLA-identical sibling donor/recipient pairs, we were able to confirm this observation and could demonstrate *NOD2/CARD15* SNPs as independent risk factors for long-term overall survival. The number of *NOD2/CARD15* variants within a donor/recipient pair clearly correlated with the likelihood to develop complications, indicating that there is a clear gene dosage effect. These important findings are now tested in a multicenter European prospective trial.

2. Materials

2.1. Isolation of Genomic DNA

1 Blood and cell culture DNA midi Kit (Qiagen, Hilden).
   a. Cell lysis buffer C1: 1.28 *M* sucrose, 40 m*M* Tris-HCl, pH 7.5, 20 m*M* $MgCl_2$, 4% Triton X-100, stored at 4°C.
   b. Digestion buffer G2: 800 m*M* guanidine HCl, 30 m*M* Tris-HCl, pH 8.0, 30 m*M* EDTA, pH 8.0, 5% Tween-20, 0.5% Triton X-100, stored at room temperature.
   c. Equilibration buffer QBT: 750 m*M* NaCl, 50 m*M* MOPS, pH 7.0, 15% isopropanol, 0.15% Triton X-100, stored at room temperature.
   d. Wash buffer QC: 1.0 *M* NaCl, 50 m*M* MOPS, pH 7.0, 15% isopropanol, stored at room temperature.
   e. Elution buffer QF: 1.25 *M* NaCl; Tris-HCl, pH 8.5; 15% isopropanol, stored at room temperature and prewarmed to 50°C before use.
2. The isolated genomic DNAs are diluted to a concentration of 20 ng/µL with sterile H_2O and stored at –20°C.

2.2. Allelic Discrimination

1. Primers (MWG Biotech, Ebersberg, Germany) are dissolved in sterile H_2O at 4 µ*M*, stored at –20°C, and used at a final concentration of 400 n*M*.

   SNP8-forward: ttc ctg gca ggg ctg ttg tc
   SNP8-reverse: agt gga agt gct tgc gga gg
   SNP12-forward: act cac tga cac tgt ctg ttg act ct
   SNP12-reverse: agc cac ctc aag ctc tgg tg
   SNP13-forward: gtc caa taa ctg cat cac cta cct ag
   SNP13-reverse: ctt acc aga ctt cca gga tgg tgt

2. Probes (MWG Biotech) are synthesised with a reporter dye 6-FAM or TET covalently linked at the 5′-end and a quencher dye TAMRA linked to the 3′-end of the probe. They are dissolved in sterile H_2O at 2.5 µ*M*, stored at –20°C in dark tubes, and used at a final concentration of 250 n*M*.

   SNP8F: FAM-cct gct ccg gcg cca ggc-Tamra
   SNP8T: TET-cct gct ctg gcg cca ggc c-Tamra
   SNP12F: FAM-ttt tca gat tct ggg gca aca gag tgg gt-Tamra
   SNP12T: TET-ttc aga ttc tgg cgc aac aga gtg ggt-Tamra
   SNP13F: FAM-cct cct gca ggc ccc ttg aaa-Tamra
   SNP13T: TET-ccc tcc tgc agg ccc ttg aaa t-Tamra

3. TaqMan® Universal PCR Master Mix (Applied Biosystems). Store at 4°C.
4. 384-Well plates (Abgene, Epsom, UK).
5. Thermocycler (Primus-HT; MWG Biotech).
6. ABI PRISM 7700 Sequence Detection System (PE Applied Biosystems, Forster City, CA).

3. Methods

Allelic discrimination by TaqMan is an easy and fast method to type DNA from multiple individuals simultaneously. Sequencing of DNA of each recipient and donor would be much more time-consuming. Thereby it is important to ensure that the analzyed DNAs have the same concentration, in order to allow better differentiation of the emitted fluorescence.

3.1. Isolation of DNA

1. Genomic DNA is isolated from EDTA-blood of recipients and donors. For this purpose, 3 mL EDTA-blood is incubated for 30 min at 50°C in a thermoblock. Three milliliters cold cell lysis buffer C1 and 9 mL cold sterile water are added, incubated for 30 min on ice, and centrifuged at 1500*g* at 4°C. After discarding the supernatant to a 1.5 mL rest volume, the pellet is resuspended by vortexing with 1 mL cold C1-buffer and 3 mL sterile water. The suspension is incubated for 10 min and centrifuged for 15 min at 1500*g*. Five milliliters digestion buffer G2 are added to the pellet, which is completely resuspended by vortexing for 30 s at maximum speed. After addition of 95 µL protease, the suspension is incubated at

50°C for 30–60 min and vortexed for 10 s at maximum speed before it is applied to the with 4 mL QBT buffer equilibrated QIAGEN Genomic-tip. The Genomic tip is washed twice with 7.5 mL of buffer QC and the genomic DNA is eluted with 5 mL of 50°C prewarmed buffer QF. DNA is precipitated by adding 3.5 mL (0.7 volumes) 100% isopropanol, inverting the tube 10–20 times, and centrifuging at 15,000*g* at 4°C for 30 min. The supernatant is carefully removed and the pellet is air-dried and resuspended with 50–200 µL sterile water. If pellet is not dissolved, store tube for some hours at 4°C and try again. If it is still undissolved, insoluble components are centrifuged and the supernatant is isolated and stored at –20°C.

2. Concentration is measured with 1:20 prediluted genomic DNA at 260 and 280 nm wavelength.

3.2. Allelic Discrimination

1. All DNAs are typed for the three major SNPs in the NOD2/CARD15 gene. A 10-µL approach for one SNP typing consists of 5 µL of 2X TaqMan Universal PCR Master Mix, 1 µL each of forward and reverse primer, 1 µL each of wild type and mutant probe, and 20 ng genomic DNA. For each SNP, a PCR-Mastermix without DNA is prepared. Nine microliters of this mastermix are transferred to a well of a 384-wellplate and 1 µL genomic DNA (20 ng) is added. Good results are also possible with DNA concentrations of 5 ng/µL. If the concentrations are lower than 5 ng/µL, it is advisable to first add the DNA, let it dry, and then add the mastermix.
2. All reactions on the 384-well plate are performed in an external thermal cycler (Primus-HT; MWG Biotech). The cycler conditions are: 50°C for 2 min, 95°C for 10 min, followed by 45 cycles of 95°C for 15 s (melting step) and 60°C for 1 min (anneal/extend step). After the PCR run, emitted fluorescence is measured by the ABI PRISM® 7700 Sequence Detection System.
3. Increases in the amount of reporter dye fluorescence during the 45 cycles of amplification are monitored using Sequence Detector software (SDS v2.2, PE Applied Biosystems). Changes in reporter dye fluorescence of 6-FAM vs changes in reporter dye fluorescence of TET are plotted (homozygous FAM vs homozygous TET vs heterozygous FAM/TET) **(Fig. 1)**.

3.3. Analysis of Data

Results of genotyping must be associated with clinical outcome variables. In the setting of stem cell transplantation (SCT), an SPSS 12.0 database was established based on the patient's charts and the typing results for *NOD2/CARD15* and used for analysis of frequencies, survival data, and risk factors. Cox regression analysis for treatment-related mortality and overall survival included *NOD2/CARD15* status, cohort and center, age, stage at the time of SCT, gender combination (female donor in male recipients vs others), intensity of conditioning (standard vs reduced intensity), stem cell source, and type

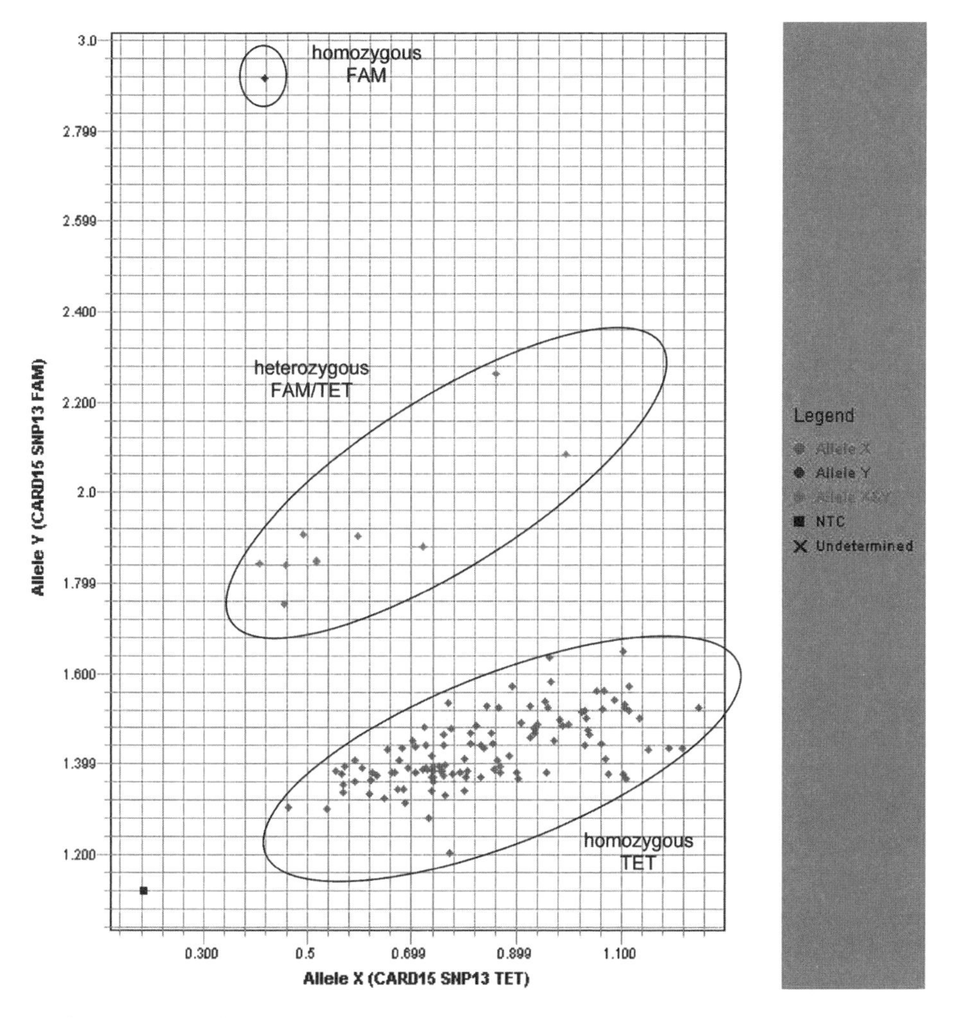

Fig. 1. Typical fluorescence plot (SNP13). The red points (homogenous TET) represent homozygous wild-type persons, the green points (heterozygous FAM/TET) persons with one mutant allele, and the blue points (homozygous FAM) people with homozygous mutant alleles.

of gastrointestinal decontamination. Analysis of the NOD2/CARD15 status had to consider the complex system of interaction between donor and recipient, as recipient SNPs may contribute to dysregulated epithelial inflammation whereas donor SNPs may affect monocyte and dendritic cell function. In our first manuscript ***(6)***, we chose the following categories:

1. Wild-type vs occurrence of any heterozygous SNP in either donor or recipient.

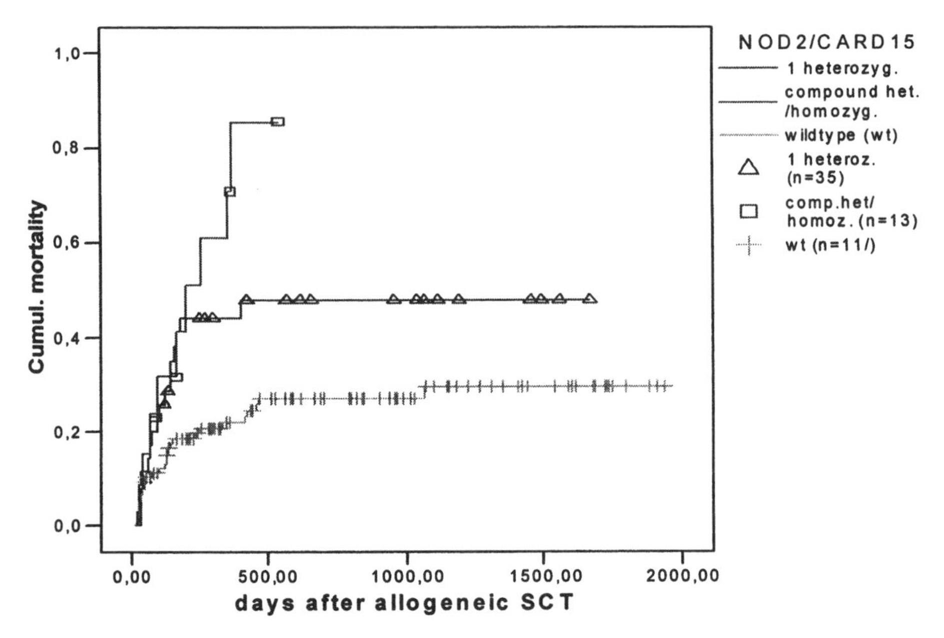

Fig. 2. Cumulative transplant-related mortality (TRM) in relation to gene dosage of NOD2/CARD15 variants in individual donor-recipients pairs. Differences were tested by log rank analysis and were 0.013 for wild-type pairs vs pairs with one heterozygous variant and <0.0001 for wild-type pairs vs the high-risk group with two or more variants.

2. Wild-type vs any variant in the recipient alone, in the donor alone, and variants occurring in both recipient and donor, which proved to be a specific high-risk group.

An alternative approach is to subgroup the *NOD2/CARD15* status according to the number of *NOD2/CARD15* variants observed in an individual donor–recipient pair: Wild-type = absence of any variant in recipient and donor; 1 variant = 1 heterozygous variant in recipient or donor; 2 or more = compound heterozygous or homozygous variants in recipients or donors or both.

Significance testing was performed by using the likelihood ratio test ($p < 0.05$). For calculation of TRM, incidence of GvHD, and relapse, the cumulative incidence method with either relapse- or nonrelapse-related mortality treated as competing risks was used with the help of the NCSS 2004 software (NCSS, Kaysville, UT). All group comparisons were done using Fisher's exact test (p values < 0.05 were considered as statistically significant). **Figure 2** gives a typical example of a gene dosage analysis performed for our first cohort of patients and donors tested for the presence of NOD2/CARD15 SNPs.

References

1. Dickinson, A. M., Middleton, P. G., Rocha, V., Gluckman, E., and Holler, E. (2004) Genetic polymorphisms predicting the outcome of bone marrow transplants. *Br. J. Haematol.***127,** 479–490.
2. Ferrara, J. L. (2002) Cellular and cytokine effectors of acute graft versus host disease. *Int. J. Hematol.* **76,** 195–198.
3. Hugot, J. P., Chamaillard, M., Zouali, H., et al. (2001) Association of NOD2 leucine-rich repeat variants with susceptibility to Crohn's disease. *Nature* **411,** 599–603.
4. Ogura, Y., Bonen, D. K., Inohara, N., et al. (2001) A frameshift mutation in NOD2 associated with susceptibility to Crohn's disease. *Nature* **411,** 603–606.
5. Hampe, J., Grebe, J., Nikolaus, S., et al. (2002) Association of NOD2 (CARD 15) genotype with clinical course of Crohn's disease: a cohort study. *Lancet* **359,** 1661–1665.
6. Holler, E., Rogler, G., Herfarth, H., et al. (2004) Both donor and recipient NOD2/CARD15 mutations associate with transplant-related mortality and GvHD following allogeneic stem cell transplantation. *Blood* **104** 889–894.

10

Role of Natural Killer Cells and Killer Immunoglobulin-Like Receptor Polymorphisms

Association of HLA and KIRs

M.Tevfik Dorak

Summary

Natural killer cells play an important role in innate immunity. They act against infected and transformed cells as part of the immune surveillance process. Their interactions with the human leukocyte antigens (HLAs) create a situation where they may act against donor hematopoietic cells following stem cell transplantation. Both killer immunoglobulin-like receptors (KIRs) and HLA types of donor and recipient are relevant in the generation of graft-vs-leukemia or graft-vs-host reactions. This chapter reviews the current knowledge on the involvement of natural killer cells in the events following hematopoietic stem cell transplantation, the structure of the genetic complex encoding the KIRs and provides a PCR-based genotyping scheme for KIR genes.

Key Words: Natural killer cells; killer Immunoglobulin-like receptors; KIRs; graft-vs-leukemia effect; polymerase chain reaction; genotyping.

1. Introduction

Natural killer (NK) cells are peripheral blood mononuclear cells (large granular lymphocytes) with an important role in nonspecific host defense. Their action is directed toward infected cells and tumor cells, but normal cells are spared. NK cells do not require activation for cytotoxicity or presentation of a peptide by major histocompatibility complex (MHC) molecules; they always have large granules of granzymes and perforin in their cytoplasm that make them constitutively cytotoxic. NK cell activity is regulated by a variety of cell surface receptors, which may be inhibitory or activating (or triggering) ***(1,2)***. In the opposing signals model of NK cell activity, the inhibitory and activating NK receptors may coexist in the same cell but the binding of inhibitory receptors by MHC class I transmits dominant inhibiting signals. This way, normal

From: *Methods in Molecular Medicine, vol. 134: Bone Marrow and Stem Cell Transplantation*
Edited by: M. Beksac © Humana Press Inc., Totowa, NJ

Table 1
Natural Killer (NK) Cell Receptors

NK receptor family	Molecular nature	Gene location	Ligands
KIR[a]	Ig-superfamily	LRC (19q13.4)	HLA-A, Bw, Cw, G
ILT/LIR	Ig-superfamily	LRC	HLA Class Ia (-G)
CD94/NKG2[b] (KLR)	C-type lectin-like	NKC (12p12.3-p13.2)	HLA Class Ib (-E)
NKG2D[c] (KLRK)	C-type lectin-like	NKC	MICA/B and MHC class I-like
NCR[c] (NKp46, NKp44, NKp30, NKp80)	Ig-superfamily	Various; including MHC, LRC	Viral hemagglutinins and others

Compiled from **refs.** ***51,65–67***.
[a] Major inhibitory receptors.
[b] NK Group 2.
[c] Major activating receptors.
KIR, killer cell immunoglobulin-like receptor; Ig, immunoglobulin; LRC, leukocyte receptor complex/cluster; HLA, human leukocyte antigen; ILT, immunoglobulin-like transcripts; LIR, leukocyte immunoglobulin-like receptor; KLR, killer cell lectin-like receptor; MHC, major histocompatibility complex.

cells expressing MHC molecules are protected. The balance between inhibitory and activating signals is important to protect autoimmunity and to preserve activity against viral infections or neoplastic transformation.

The main target for NK cells is a cell that is missing the self-MHC class I molecules to act as an inhibitory signal for cytotoxicity *(3)*. As suggested by the activity against cells devoid of self-MHC molecules, the MHC class I molecules (human leukocyte antigen [HLA]-A, -B, and -Cw) are ligands for the inhibitory NK cell surface receptors. Besides these, MHC class I-like molecules (HLA-E, MICA, MICB, and others) also interact with different types of NK cell surface receptors, generally with activating receptors (**Table 1**). These MHC class I molecules act as indicators of overall MHC class I expression (HLA-E) or as stress signals (MICA, MICB, and HSP60) *(4,5)*. The only exception to the dominance of inhibitory signals is the activating killer cell lectin-like receptor (KLR) K (originally NK Group 2 Member D; NKG2D) which can override inhibitory signals when engaged with its ligands MICA, MICB, and other MHC class I-like molecules such as retinoic acid early inducible (RAE)-1, H60 minor histocompatibility molecules, and CMV UL-binding proteins (ULBP) *(4)*, although this stimulatory signal generated by NKG2D is not entirely refractory to inhibitory signals ***(6)***.

1.1. Killer Cell Immunoglobulin-Like Receptors

The inhibitory receptors, specific for MHC class I molecules, allow NK cells to discriminate between normal cells and cells that have lost the expression of MHC class I (e.g., tumor cells). The major group of inhibitory NK cell receptors are killer cell immunoglobulin-like receptors (KIRs), which were originally called killer inhibitory receptors. The peptides within the MHC class I cleft do not take part in the interaction with NK receptors; however, certain side chains at position 7 and 8 of the nonamer peptide interfere with binding of certain KIRs (KIR2DL and KIR3DL) ***(7)***. There is also some evidence that activating receptors specific for the same HLA may respond differentially depending on bound peptides ***(8)***. KIRs generally interact with MHC class I molecules but an exception involving KIR2DS4 has recently been reported. KIR2DS4 interacts with HLA-Cw4 for activation but this is not a strict requirement, as other non-MHC ligands can still bind to this KIR with functional consequences ***(9)***.

KIR molecules are encoded in the leukocyte receptor complex/cluster (LRC) on human chromosome 19q13.4, which spans approx 1 Mb ***(10,11)***. The LRC is polygenic and individual genes exhibit polymorphism ***(12–14)***. This region is flanked by Fc alpha receptor (CD89), immunoglobulin (Ig)-like transcripts (ILT; including CD85, also called leukocyte immunoglobulin-like receptor [LIR]) and monocyte-macrophage inhibitory receptor (MIR) gene families ***(11)***. The ILTs are also inhibitory receptors using HLA class I as ligands. They are expressed on monocytic cells, dendritic cells, and some NK and B cells. The KIR genes do not undergo somatic rearrangement (unlike TCR or Ig genes) but the number of genes (especially the noninhibitory ones) on each haplotype is variable ***(12,13)***. More than 100 highly homologous KIR variant sequences have been deposited in databases and more sequences are reported as different ethnic groups are examined ***(15–19)***. Therefore, the KIR genetic repertoire is characterized by variable gene content and allelic polymorphism resulting in a probability of <0.01 for two unrelated individuals to have the same KIR genotype ***(13,14)***. Different clones within an individual may each express a unique subset of the available KIR repertoire ***(15,20)***.

1.2. KIR Gene Content of the LRC

Within the LRC, different broad gene families can be identified by phylogenetic analysis, number of extracellular Ig domains (2D or 3D) and length of cytoplasmic tail (S or L). While those with long (L) cytoplasmic tails encode inhibitory receptors (2DL and 3DL groups), the activating ones have short (S) cytoplasmic tails (2DS and 3DS groups). The gene KIR2DL4 (p49, CD158d) is

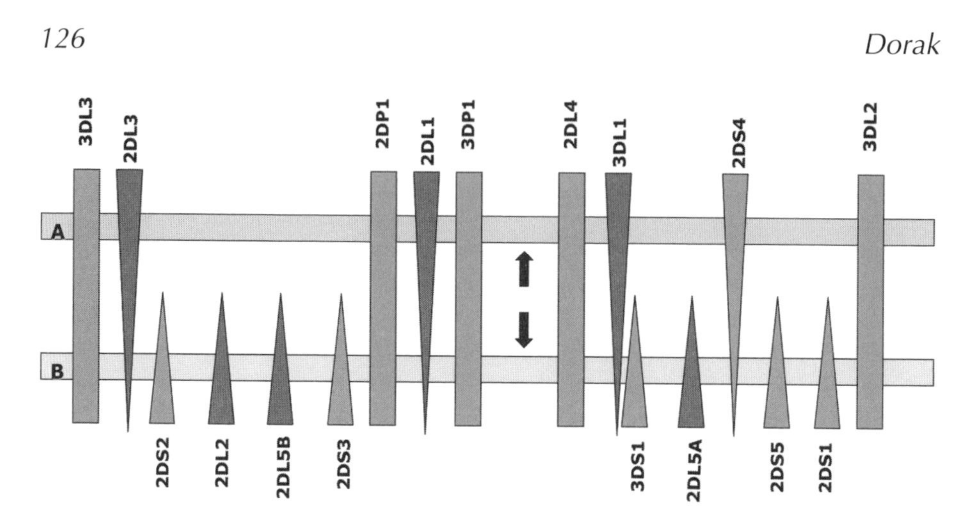

Fig. 1. Haplotypic distribution of 17 KIR genes.

present on all haplotypes (a framework gene) and almost ubiquitously expressed *(21,22)*. Its ligand may be HLA-G *(12,21)*. Despite having a long cytoplasmic tail, KIR2DL4 exhibits activating function but also has inhibitory potential *(23)*. Each KIR group consists of one to five members that differ by 1–9 nucleotide substitutions, whereas members of different subfamilies differ by at least 20 nucleotides. The 17 KIR genes (of which 2 are pseudogenes) officially recognized by the KIR Nomenclature Committee *(24)* are shown in **Table 2.** Also shown in **Table 2** are the CD designations as members of the CD158 series given to KIR proteins.

Two broad groups, A and B, have been proposed for segregating KIR haplotypes in human populations (**Fig. 1**) *(13,15)*. Each of these haplotypes may have different number of KIR genes in different individuals. Haplotype A comprises of KIR genes 2DL1, 2DL3, 3DL1, and 2DS4 that occur on haplotype A more frequently than on haplotype B. Haplotype B is characterized by one or more of the following genes that occur exclusively on this haplotype: 2DS1, 2DS2-2DL2 (co-segregation), 2DS3, 2DS5, 2DL2, 2DL5, and 3DS1 *(2,24,25)*. Among these, the most characteristic members of each haplotype are 2DL3 for haplotype A, and 2DL2 and 2DS2 for haplotype B (**15–17**,**26**). There is no haplotype A-specific gene that does not occur on haplotype B. Even 2DL3, which is present on all A haplotypes, also occurs on about 8% of Caucasian haplotype B group *(2)*. Because of this, the current practice of haplotype assignment involves exclusion of haplotype B to assign a haplotype as haplotype A. Thus, a haplotype that carries one or more of 2DL3, 2DL1, 3DL1, or 2DS4, but none of the haplotype B-specific genes, is assigned as haplotype A. KIR2DS4

Table 2
Killer Cell Immunoglobulin-Like Receptor (KIR) Genes

Symbol	Full name	Aliases and CD designation	Number of alleles	Function
KIR2DL1	killer cell immunoglobulin-like receptor, two domains, long cytoplasmic tail, 1	NKAT1, p58.1, CD158a	6	Inhibitory
KIR2DL2	killer cell immunoglobulin-like receptor, two domains, long cytoplasmic tail, 2	NKAT6, CD158b1	5	Inhibitory
KIR2DL3	killer cell immunoglobulin-like receptor, two domains, long cytoplasmic tail, 3	NKAT2, NKAT2a, NKAT2b, p58, CD158b2	6	Inhibitory
KIR2DL4[a]	killer cell immunoglobulin-like receptor, two domains, long cytoplasmic tail, 4	p49, CD158d	9	Inhibitory/ Activating
KIR2DL5A	killer cell immunoglobulin-like receptor, two domains, long cytoplasmic tail, 5A	KIR2DL5.1, CD158f	3	Inhibitory
KIR2DL5B	killer cell immunoglobulin-like receptor, two domains, long cytoplasmic tail, 5B	KIR2DL5.2, KIR2DL5.3, KIR2DL5.4	2	Inhibitory
KIR2DS1	killer cell immunoglobulin-like receptor, two domains, short cytoplasmic tail, 1	CD158h	4	Activating
KIR2DS2	killer cell immunoglobulin-like receptor, two domains, short cytoplasmic tail, 2	NKAT5, CD158j	5	Activating
KIR2DS3	killer cell immunoglobulin-like receptor, two domains, short cytoplasmic tail, 3	NKAT7	3	Activating

(Continued on next page)

Table 2 ***(Continued)***
Killer Cell Immunoglobulin-Like Receptor (KIR) Genes

Symbol	Full name	Aliases and CD designation	Number of alleles	Function
KIR2DS4	killer cell immunoglobulin-like receptor, two domains, short cytoplasmic tail, 4	NKAT8, CD158i	6	Activating
KIR2DS5	killer cell immunoglobulin-like receptor, two domains, short cytoplasmic tail, 5	NKAT9, CD158g	2	Activating
KIR3DL1	killer cell immunoglobulin-like receptor, three domains, long cytoplasmic tail, 1	NKAT3, CD158e1	11	Inhibitory
KIR3DL2[a]	killer cell immunoglobulin-like receptor, three domains, long cytoplasmic tail, 2	NKAT4, NKAT4a, NKAT4b, CD158k	12	Inhibitory
KIR3DL3[a]	killer cell immunoglobulin-like receptor, three domains, long cytoplasmic tail, 3	KIRC1, KIR3DL7, KIR44, CD158z	1	Inhibitory
KIR3DS1	killer cell immunoglobulin-like receptor, three domains, short cytoplasmic tail, 1	NKAT10, CD158e2	4	Activating
KIR2DP1[a]	killer cell immunoglobulin-like receptor, two domains, pseudogene 1	KIRZ, KIRY, KIR2DL6	1	Pseudogene
KIR3DP1[a]	killer cell immunoglobulin-like receptor, three domains, pseudogene 1	KIRX, KIR2DS6, KIR3DS2P, CD158c	2	Pseudogene

Adapted from http://www.gene.ucl.ac.uk/nomenclature/genefamily/kir.html and http://www.ebi.ac.uk/ipd/kir/loci.html (*see also* **refs. *1,24,29***).
[a] Framework (anchor) genes present on all KIR haplotypes.

is the only activating receptor on haplotype A, although it is present on more than 70% of B haplotypes. It usually occurs as a nonfunctional 22-bp deletion variant (KIR1D) in Caucasians *(25,27)*, but not on Korean KIR haplotypes. The KIR genes 2DL4, 3DP1, 3DL2, and 3DL3 are present on almost all haplotypes and called framework or anchor genes *(10,25,28)*. The previously suggested haplotype C is no longer accepted as the ability to type a then unknown variant of KIR2DL1 (KIR2DL1v) identified that haplotype as a member of haplotype A group *(17,19)*.

Haplotype A is usually the more common one but the frequencies of the haplotypes vary considerably among the ethnic groups *(16–19,25)*. Haplotype A has now been subgrouped into two different genotype groups: (1) A-1D containing KIR2DS4 and (2) A-2DS4 containing KIR2DS4*003 (KIR1D) *(25)*. The haplotype A-1D lacks all functional 2DS-activating KIRs and only contains the central framework gene KIR2DL4 as the sole receptor on this haplotype with activating function. Several pairs of KIR genes and pseudogenes display negative linkage disequilibrium, implying they could be alleles. Thus, they do not occur on the same haplotypes. These are: 2DL2/2DL3, 3DS1/3DL1, 2DS3/2DS5, and 2DS1/2DS4/ 2DS4*003 (KIR1D) *(28)*. Supported by high-resolution and linkage disequilibria studies, the following assumptions are usually made in haplotype determination studies *(13,25)*: all haplotypes contain the framework genes 3DL3, 2DL4, and 3DL2; haplotypes contain either 2DL2 or 2DL3/2DL1, but not 2DL2 and 2DL3/2DL1; and haplotypes contain either 3DP1 or 3DP1 variant (3DP1v), but not both; 3DS1 segregates as an allele of 3DL1; 2DS4 is present on haplotypes that do not have any other 2DS genes. In addition, 2DS4-containing haplotypes contain either one of the allelic variants 2DS4 deletion variant (KIR1D) or 2DS4 *(27)* but not both. This simplified haplotype A and B scheme has been refined and all haplotypes are now thought of a combination of one of five centromeric half variants and one of two telomeric half variants *(28)*. Some rare haplotypes may still occur as recombinants. The group A KIR haplotypes show higher allelic diversity than B haplotypes. The latter is characterized by gene content diversity with minor allelic polymorphisms.

Because haplotype B is rich in activating receptors as opposed to the lack of them on haplotype A *(25,27,28)*, individuals who are homozygous for group B haplotypes will have qualitatively greater potential NK cell activation than group A homozygotes *(14)*.

1.3. Allelic Polymorphisms of KIR Genes

Although KIR genes show presence or absence variation on haplotypes, each gene may show further allelic polymorphism *(13,25,28,29)*. From 17 KIR genes, 89 alleles have been unambiguously defined and fully sequenced *(30)*.

Since the Nomenclature Committee publication, a further 20 alleles had been named with more pending *(31)*. The sequences are available at the Immuno-Polymorphism Database *(32)* and any new sequence can be compared with the available sequences using the online dbLRC resource *(33)* (**Note 1**).

KIR allele sequences are named in an analogous fashion to HLA allele nomenclature. After the gene name, an asterisk is used as a separator before a numerical allele designation. The first three digits of the numerical designation are used to indicate alleles that differ in the sequences of their encoded proteins. The next two digits are used to distinguish alleles that only differ by synonymous (noncoding) differences within the coding sequence. The final two digits are used to distinguish alleles that only differ by substitutions in an intron, promoter, or other noncoding region of the sequence *(24)*.

1.4. KIR Expression

The KIR genes have a unique and fascinating expression pattern. Most KIR loci are expressed on different NK clones in a stochastic manner. This diversity is achieved through stochastic expression of genes from the genomic repertoire in individual clones rather than the combinatorial receptors characteristic of T-cell receptor and immunoglobulins in the lymphoblasts. Within an individual, each NK cell expresses two to nine different KIRs in different combinations along with the CD94/NKG2 heterodimer *(34)*. To ensure self-tolerance, normally at least one inhibitory KIR specific for a host HLA class I allele is expressed by any NK cell although not necessarily for all allogeneic HLA class I *(20,34,35)*; if not, this deficit is compensated by the expression of CD94: NKG2A *(34,35)*. This is the reason for the inverse correlation between the expression levels of KIRs and CD94:NKG2A *(35)*. The expression of a KIR is determined at the transcription level and is stable in the progeny of NK cells *(34–36)*. This pattern of expression is regulated by methylation of KIR loci and may also result in allele-specific expression *(22,37,38)*. Allelic polymorphism of KIR loci may correlate with expression levels. Differences in the levels of KIR3DL1 expression are due to allelic polymorphism in the KIR3DL1 gene *(39)*. KIR2DL4 is polymorphic, and some common alleles, although transcribed, do not produce cell surface KIR2DL4 *(23)*.

Random expression of individual KIR receptor genes is a rule with some exceptions. Significant increases in the frequencies of cells expressing the combinations 2DL1/2DS1, 2DL2/2DS2, and 2DS1/2DS2 over those expected from the product of their individual frequencies have been reported *(20)*. This suggests the presence of a nonstochastic component in the regulation of expression in addition to the generally stochastic nature. The molecular mechanisms that regulate the clonally diverse expression of KIR genes on NK and T cells are not known *(5)*.

The expression of KIRs is not regulated by self-MHC and is not inherited in an HLA-linked manner *(20,36)*. Individuals may have any combination of KIRs regardless of their HLA type even though the KIRs they have may not have the correct ligands for them in the HLA type of the individual *(40)*. The impact of HLA is to change the frequencies of KIR-expressing cells, although they have no effect on the surface levels of KIR expression *(35)*. The HLA class I genotype dictates the number of KIRs that can serve as inhibitory receptors for autologous HLA class I, and thus the proportion of NK cells needing CD94: NKG2A expression to be tolerant of self. In mice, however, the expression of the Ly49 group of (inhibitory) NK receptors seems to be regulated by the MHC *(41)*.

A study that evaluated KIR expression by flow cytometry, reverse-transcription (RT)-PCR, and quantitative real-time PCR has shown that expression levels of KIR genes are highly variable among individuals. Only KIR2DL3 and KIR3DL2 are always expressed in all individuals and others show variation including no expression *(38)*. That study identified KIR2DL1 and KIR3DL1 as the ones expressed by less than 15% of the carriers of these genes. Subjects who possess both KIR2DL2 and KIR2DL3 appear to preferentially express KIR2DL3. Leung et al. also identified epigenetic silencing of KIR2DL1 in their study of KIR expression *(38)*. These findings have strong implications in interpretation of KIR genotyping results and may even explain some of the inconsistencies in reported associations of KIR genotypes with hematopoietic stem cell transplantation (HSCT) outcome. In fact, if KIR typing is going to be used for clinical decisions, both genotyping and phenotyping should be performed *(38)*.

Expression of KIRs is not exclusive to NK cells; a subset of γδTCR+ T-cells and memory/effector αβTCR+ T cells also express KIRs *(1)*. A KIR-expressing subpopulation of T-lymphocytes consist of less than 2% of CD3+ T-cells and displays a cell surface phenotype typical of memory CD8+ T-cells (CD45RO+ CD29+CD28-CD45RA-). One subtype of T-cells, CD4(+)CD28(null) T-cells, are a highly oligoclonal subset of T-cells that is expanded in patients with rheumatoid arthritis. In CD8+ self-reactive T-cells, TCR engagement sustains KIR expression *(42)*. It is believed that KIR expression may mediate T-cell tolerance to self-antigens by sparing self-reactive T-cells.

1.5. KIR Ligands

Most inhibitory KIRs have well defined MHC class I ligands (**Tables 3** and **4**) whereas the ligands for activating KIRs have not been characterized that well *(1,2)*. It is possible that activating KIR ligands are non-HLA molecules as suggested by weak binding of activating KIRs to HLA class I *(43)*. Recently, an example of this has been reported *(9)*.

Table 3
Killer Cell Immunoglobulin-Like Receptor (KIR) Ligand Specificity

2DL1 and 2DS1	2DL2/3 and 2DS2	3DL1/3DS1	3DL2	2DL4	2DS4
HLA-C2 group	HLA-C1 group	HLA-B - Bw4	HLA-A	HLA-G	HLA-C
C*02	C*01	B*08	A*03		C*04
C*04	C*03	B*13	A*11		
C*05	C*07	B*27			
C*06	C*08	B*44			
		B*51			
		B*52			
		B*53			
		B*57			
		B*58			

Ligands for 2DL5, 2DS3, 2DS5, and 3DL3 are unknown.

The combination of homozygosity for KIR haplotype B and homozygosity for HLA-C2 epitope creates a situation where no KIR will have an inhibitory ligand and NK cell inhibition will rely mainly on the CD94:NKG2A receptor. This combination may occur following HSCT when both the recipient's HLA-C alleles belong to the HLA-C2 group (Lys80-bearing alleles, e.g., Cw*02, *04, *05, *06, *15, *1602, *17, *18) and the donor's KIR repertoire consists of haplotype B (mainly activating receptors) only. This would be a favourable KIR–HLA interaction in terms of NK cell activity against leukemia cells (graft-vs-leukemia [GvL] effect).

A hierarchy of the strength of inhibition or activation for different KIR–HLA ligand combinations has been recognized ***(2,8,44,45)***. For example, the inhibitory signal generated by KIR2DL1-HLA-C2 is the strongest. Seemingly paradoxical associations reported for hepatitis C virus (HCV) infection ***(44)*** and human papilloma virus (HPV)-induced cervical cancer ***(45)*** have been explained by this relationship.

1.6. KIRs and HSCT

Currently, the preferred transplant donor is an HLA-identical sibling or an HLA-matched unrelated individual. Because HLA and KIR loci are not linked, even HLA-identical siblings have some degree of KIR mismatching ***(35,46)***. After engraftment, the initial NK cell populations express CD94:NKG2A at a higher level than KIRs. Gradually, the frequency of CD94:NKG2A decreases and KIR expression increases ***(46)***. Eventually, KIR expression resembles that of the donor and can be different what the recipient had prior to transplantation ***(47)***.

Unlike allogeneic T-cells, NK cells can discriminate tumor cells from normal healthy tissues. Thus, they have the capacity to mount a GvL effect without

Table 4
C1 / C2 and Bw4/6 Epitopes as Killer Cell Immunoglobulin-Like Receptor (KIR) Ligands

A. HLA-Cw alleles with C1 epitope and HLA-B alleles in linkage disequilibrium[a,b]

HLA-Cw	HLA-B	HLA-B epitope
Cw*03	B*55 (Cw*09)	Bw6
	B*62 (Cw*09/10)	Bw6
	B*60 (Cw*10)	Bw6
Cw*07	B*08	Bw6
	B*07	Bw6
	B*18	Bw6
	B*39	Bw6
Cw*08	B*14	Bw6
Cw*12		
Cw*13		
Cw*14		
Cw*1601		

B. HLA-Cw alleles with C2 epitope and HLA-B alleles in linkage disequilibrium[a,b]

HLA-Cw	HLA-B	HLA-B epitope
Cw*02	B*27 (Cw*01/02)	Bw4
	B*61	Bw6
Cw*04	B*35	Bw6
	B*53	Bw4
Cw*05	B*44	Bw4
	B*18	Bw6
Cw*06	B*13	Bw4
	B*37	Bw4
	B*45	Bw6
	B*57	Bw4
	B*47	Bw4
	B*50	Bw6
Cw*15		
Cw*1602		
Cw*1701		
Cw*18		

[a] In Caucasians.
[b] The C1 and C2 are epitopes correspond to Casn80 and Clys80, respectively.

causing graft-vs-host-disease (GvHD). An earlier observation that matching at HLA-C was a significant risk factor for leukemia relapse after transplantation was interpreted as HLA-C mismatch may elicit GvL effect ***(48)***. This observation suggested the involvement of NK cells in this phenomenon because of the role played by HLA-C molecules in NK cell biology. Initial studies in haploidentical donor HSCT (high-dose stem cells and T-cell depletion in the conditioning regimen with no posttransplantation immunosuppression for GvHD prevention) indeed reported on a favorable effect of KIR mismatching on leukemia relapse and other clinical outcomes (graft failure, GvHD) ***(49–51)***. The analysis of recipient–donor pairs mismatched at KIR ligand specificities (HLA-B and HLA-C) showed that when the inhibition of donor NK cells was predicted to be lifted (because of the absence of at least one of the inhibitory ligands for donor NK cell clones in the recipients) there was an enhanced anti-leukemic effect of NK cells in vitro. As an in vivo correlation, these patients had a lower risk of relapse. These studies suggested that KIR-specific epitope mismatching may be a useful strategy for enhancing the GvL effect of allogeneic haploidentical donor HSCT. Although the first study, which investigated this phenomenon in unrelated donor HSCT (without T-cell depletion) did not confirm it ***(52)***, another one that used antithymocyte globulin (ATG) in the conditioning regimen showed the same effect ***(53)***. More recent studies continue to support a contribution of HLA mismatches to better survival in AML as a result of an interaction with KIRs in unrelated donor HSCT ***(38,54–56)***. Discrepancies seem to have been due to differences in the transplantation settings used at different treatment centers (such as conditioning regimens, stem cell and T-cell content, and the different forms of GvHD prophylaxis) and the heterogeneity in leukemic group (alloreactive NK cells offer no protection from ALL following haploidentical transplantation ***[49]***). The overall results predict that the presence of inhibitory KIRs on the unrelated donor's NK cells and the absence of corresponding KIR ligand in an AML patient's (recipient) HLA repertoire ("receptor–ligand" or "missing KIR ligand" model) may be used in clinical decisions.

Algorithms have been presented for determination of clinically relevant recipient HLA and donor KIR combinations ***(38,55)***. If the recipient's HLA-C genotype consists of alleles with Asn80 only (C1 group: Cw*01, 03, *07, *08, *12, *13, *14, *1601/4), donors with inhibitory KIRs using the C2 group as ligands (2DL1 and 2DL2/2DL3) will create a "receptor–ligand mismatch" after transplantation (2DL2/2DL3 will not matter because they most likely use both C1 and C2 epitopes as ligands). Likewise, for recipients lacking HLA-Bw4, presence of KIR3DL1 will result in a receptor–ligand mismatch for an inhibitory KIR. These specific combinations are crucial because of the dominance of inhibitory signals in NK cell cytotoxicity. The donor's HLA type plays no role in this interaction. Because of the stochastic nature of KIR expression, either combination is relevant regardless of the other. There may be NK cell clones that express KIR2DL1 as the only KIR and whether the recipient has HLA-Bw4 or not does not matter for that clone. This clone will see the recipient cells as missing the ligand for its KIR. Because of the inverse correlation of KIRs and CD94:NKG2A expressions *(35)*, the more universal NKG2A - HLA-E interaction may not be enough to prevent cytotoxicity for this clone. The GvL effect, which occurs when the donor cells express an inhibitory KIR in the absence of its corresponding ligand in the recipient, results in lower relapse rates if certain transplant conditions are met. This is the basis of reported favorable effects of "KIR mismatching" following HSCT. Leung et al. cautioned that two of the most important KIRs, 2DL1 and 3DL1, have variable expression levels, and genotyping alone may not be able to predict their presence on the donor NK cells *(38)*.

1.7. KIR Genotyping

An individual's KIR genotype can be determined by flow cytometry for surface expression *(38,39,46,56,57)*, by RT-PCR *(15,20,38)* or real-time PCR *(38)* at the RNA level, or by a variety of methods at the DNA level. Recently, a matrix-assisted laser/desorption ionization (MALDI)-time-of-flight (TOF) mass spectrometer method using primer extension has also been used *(31)*. Because of similar levels of polymorphism, KIR genotyping shares same principles of HLA typing (*see* Chapter 3). As in HLA typing, the main methods have utilized sequence-specific primers (SSP) ***(13,15,17,25,58–60)***, sequence-specific oligonucleotides (SSOs) ***(26,60,61)***, single-stranded conformation polymorphism (SSCPs) ***(62)***, or sequencing-based typing (SBT) ***(30,33)***. Most recently, a multiplex PCR-SSP method has also been described ***(63)***. Among all of these methods, PCR-SSP is the most widely used and is able to cover all known alleles. The currently used PCR-SSP methods derive from the first scheme described by Uhrberg et al. ***(15)*** to account for newly discovered loci, previously undetected alleles, and for variants affecting primer annealing ***(13,17,25,58–60)***. PCR-SSP is also the method of choice by the International Histocompatibility Working Group NK Receptors and HLA Polymorphism Project (http://www.ihwg.org/components/nkover.htm) (**Note 2**). A medium resolution PCR-SSP typing kit has been produced by Pel-Freez/Dynal. This kit detects the presence or absence of all 17 KIR genes and also provides limited information on common variants of 2DL5, 2DS4, and 3DP1 using 21 PCR amplification reactions. Most methods use the framework genes (2DL4, 3DP1, 3DL2, and 3DL3) that are present on all haplotypes as internal controls. Among those, 2DL4 and 3DL2 are present on all haplotypes in all ethnic groups examined to date ***(19)*** and thus the most suitable genes to be used as positive internal controls in PCR-SSP methods (**Notes 3** and **4**). RT-PCR based on PCR-SSP is the method of choice for allotyping NK cell clones and remains largely unchanged from that described previously ***(15)***.

By genotyping at the DNA level, an overall impression of the KIR repertoire is obtained. PCR-SSO ***(26,60)***, PCR-SSP ***(15,17,19,25,59,60)***, and multiplex PCR-SSP *(63)* have been established as typing techniques. The minimum requirements are 12 primer pairs for PCR-SSP or 13 probes for PCR-SSO. These techniques however only address to the variable gene content of KIR haplotypes but do not deal with allelic polymorphism that has recently been recognized *(13)*. Sequence-based typing has not been used routinely to type KIR loci or alleles but dbLRC (**Notes 1** and **5**) has an alignment viewer for analyzing KIR sequence data *(33)* (**Note 6**).

Currently, there is no single typing scheme that is useful for typing all 16 genes and relevant alleles of few of them. A combination of the most up-to-date schemes *(19,25,58,59)* may be needed for the most comprehensive analysis of the KIR gene content. As mentioned previously, KIR genotyping results should be correlated with protein expression studies (flow cytometry) for predictive use in HSCT setting. KIR typing at the protein level is not covered in this chapter (*see* **refs.** ***38,39,46,56,57***).

2. Materials

High-molecular-weight genomic DNA, standard PCR reagents, thermal cycler, and visualization methods are needed as in any other genotyping method. Because of relatively large amplicon sizes for certain genes, integrity of the DNA samples is particularly important. It is important to choose a large size internal control to ensure the integrity of the DNA (**Note 4**). For the same reason, it may be desirable to use a high-performance Taq polymerase rather than a regular one. The genotyping procedure does not differ from the other typings of polymorphic systems such as the HLA alleles. The main difference is the sequence-specific primers. A typical genotyping scheme requires the following:

1. Genomic DNA. Any method that provides high-molecular-weight, stable and undegraded DNA is suitable. This is best achieved using widely available DNA extraction kits (e.g., kits by Qiagen, Nucleon, Promega, Gentra) but a manual phenol-chloroform extraction method can also be used. Quick extraction methods should be avoided in the interest of stability of the DNA during long-term storage. Ideal sources for genomic DNA are peripheral blood, bone marrow, or buccal cell samples. Archived samples such as paraffin-embedded tissue, stained blood smears, or dried blood spots will provide genomic DNA, but the DNA obtained from these sources may not be suitable for amplification of a relatively long amplicon as is required for KIR genotyping (**Note 7**). The method being used should yield long DNA fragments (such as Amersham or Qiagen kits). The amount of genomic DNA required varies between 10 and 200 ng depending on the reaction volume, material used, and the thermal cycler. Smaller reaction volumes in thin-walled tubes or plates, high-performance Taq polymerase, and most recent versions of thermal cyclers require the smallest amount of DNA. Original KIR genotyping was performed using cDNA, and most methods are suitable for KIR genotyping on cDNA samples.
2. Primers. The primers used in a typical KIR genotyping are presented in **Table 5.** Most of those are from Gagne et al. *(58)* because of shorter amplicon sizes and

Table 5
A List of PCR-SSP Primers for Detection of the Presence of KIR Genes

KIR	Forward primer (59–39)	Reverse Primer (59–39)	Size (bp)
2DL1[a]	GTTGGTCAGATGTCATGTTTGAA	GGTCACTGGGAGCTGACAC	204
	TGGACCAAGAGTCTGCAGGA	ACTCAGCATTTGGAAGTTCCG	270
2DL2[a]	CTGGCCCACCCAGGTCG	GGACCGATGGAGAAGTTGGCT	173
	TCATCCTCTTCATCCTCCTCT	AGCATTTGGAAGTTCCGC	750
2DL3[a]	CTTCATCGCTGGTGCTG	AGGCTCTTGGTCCATTACAA	550
	TCCTTCATCGCTGGTGCTG	GGCAGGAGACAACTTTGGATCA	800
2DL4[a]	CAGGACAAGCCCTTCTGC	CTGGGTGCCGACCACT	254
	ACCTTCGCTTACAGCCCG	GGGTTTCCTGTGACAGAAACAG	288
3DL1[a]	CCAGCGCTGTGGTGCCTCGA	TGGGTGTGAACCCCGACATG	197
	GGTGAAATCAGGAGAGAG	GTAGGTCCCTGCAAGGGGAA	181
3DL2[a]	CAAACCCTTCCTGTCTGCCC	GTGCCGACCACCCAGTGA	245
	CCCATGAACGTAGGCTCCG	CACACGCAGGGCAGGG	130
3DL3[a]	GTCAGATGTCAGGTTTGAGCG	CATGGAATAGTTGACCTGGGAAC	112
	GCAGCTCCCGGAGCTTG	GGGTCTGACCACGCGTG	190
2DL5[b]	TGCCTCGAGGAGGACAT	CCGGCTGGGCTGAGAGT	1151
2DL5.1[c]	CTCCCGTGATGTGGTCAACATGTAAA	GGGGTCACAGGGCCCATGAGGAT	1883
2DL5.2[c]	GTACGTCACCCTCCCATGATGTA	GGGGTCACAGGGCCCATGAGGAT	1893

(Continued on next page)

Table 5 ***(Continued)***
A List of PCR-SSP Primers for Detection of the Presence of KIR Genes

KIR	Forward primer (59–39)	Reverse Primer (59–39)	Size (bp)
2DS1[d]	GTAGGCTCCCTGCAGGGA	CAGGGCCCAGAGGAAAGTT	149
	CTTCTCCATCAGTCGCATGAA/G	GACAAACAAGCAGTGGGTCACTT	184
2DS5[d]	GCACAGAGAGGGGACGTTTAACC	ATGTCCAGAGGGTCACTGGGC	179
	CTCCCCCTATCAGTTGTCAGCG	AAAGAGCCGAAGCATCTGTAGGC	1816
2DS2[a]	TTCTGCACAGAGAGGGGAAGTA	AGGTCACTGGGAGCTGACAA	173
	CGGGCCCCACGGTTT	GGTCACTCGAGTTTGACCACTCA	240
2DS3[a]	TGGCCCACCCAGGTCG	TGAAAACTGATAGGGGGAGTGAGG	242
	CTATGACATGTACCATCTATCCAC	AAGCAGTGGGTCACTTGAC	190
2DS4[a]	CTGGCCCTCCCAGGTCA	TCTGTAGGTTCCTGCAAGGACAG	204
	TAGGCTCCCTGCAGTGCG	GAGTTTGACCACTCGTAGGGAGC	129
3DS1[a]	GGTCCAGAGGGCCGGT	GAAGAGTCACGTGTCCTCCG	900
	GGTGAAATCAGGAGAGAG	GTCCCTGCAAGGGCAC	177
2DP1[c]	TCTGTTACTCACTCCCCCA	GGAAAGAGCCGAAGCATC	1825
3DP1[b]	CCATGTCGCTCATGGTCG	TGACCACCCAGTGAGGA	361/1834
	CCATGTCGCTCATGGTCG	AGGGTTCTTCTTGCTGC	1047

[a] From **ref. *58***.
[b] From **ref. *59***.
[c] From **ref. *25***.
[d] From **ref. *19***.
PCR-SSP, PCR using sequence-specific primers; KIR, killer cell immunoglobulin-like receptor.

also because their scheme uses two primer sets for 11 of the expressed genes to allow duplicate testing of each gene in case yet-unknown polymorphisms cause loss of function of one of the primers. The longest amplicon is 900 bp (for KIR3DS1). The rest of the duplicate primer sets for the expressed genes (for 2DS1 and 2DS5) are from Cook et al. ***(19)***. The primers for KIR2DL5 (A and B) and its alleles and for the two pseudogenes KIR2DP1 and KIR3DP1 are from Gomez-Lozano and Vilches ***(59)***. KIR2DS4 variants 2DS4F and 2DS4D can be typed by primers described by Hsu et al. ***(25)***.

3. Internal control. Any of the framework genes that are present in each sample may be considered as internal control for the PCR reaction but an unrelated gene may also be included in the scheme as internal control. Whatever gene is being used as internal control, its primers must be included in each reaction alongside the primers for the target gene. This multiplexing is routine for any PCR-SSP typing scheme and requires some optimization.

3. Methods

PCR conditions do not differ from standard PCR-SSP conditions. Briefly, following optimization, primers, dinucleotide mix (dNTPs), and reaction buffer are mixed and preferably a high-performance Taq polymerase is added to each tube or well. It is critical that all general laboratory safeguards are in place to avoid contamination. The following protocol is from Gagne et al. ***(58)*** and provides general conditions. It is, however, best to optimize the primer concentrations and cycling parameters locally for the reagents and equipment used. Because of the long amplicon sizes, it may be useful to keep extension times a little longer than usual.

KIR-specific primers are used at a final concentration of 0.1 to 0.5 μ*M* depending on the target gene. Amplifications of KIR genes are performed in 10-μL reactions using 0.4 U Taq polymerase, 0.2 m*M* of each dNTP, 1.5 m*M* $MgCl_2$, 5% glycerol, and 50 ng of genomic DNA under the following conditions: initial denaturation for 2 min at 92°C, then 25 s at 91°C, 45 s at 65°C, 30 s at 72°C for four cycles, 25 s at 91°C, 45 s at 60°C, 30 s at 72°C for 26 cycles, 25 s at 91°C, 60 s at 55°C, 120 s at 72°C for 5 cycles, and 10 min at 72°C. Amplification products are visualized on a 2% agarose gel following staining with ethidium-bromide (or other DNA binding dyes). It is important to keep an image of the gel or take a picture for permanent records.

4. Notes

1. IPD-KIR Database (http://www.ebi.ac.uk/ipd/kir) and dbLRC Resource (http://www.ncbi.nlm.nih.gov/mhc/MHC.fcgi?cmd=init&locus_group=2)
2. For IHWG reference cell lines typing results, *see* **refs.** ***25,59***, and ***60***.
3. The framework genes 2DL4, 2DP1, 3DP1, 3DL2, and 3DL3 that are present on all haplotypes can be used as internal controls for PCR-based detection methods.

4. If setting up a new SSP method, for the selection of an internal control, it is best to consider a framework gene whose amplicon will be larger than any of the target genes. This way, when present, the target gene will be amplified preferentially.
5. The KIR3DL2*001 sequence is the reference KIR sequence as it provides a sufficiently long reference sequence and also possesses a high level of nucleotide identity and structural homology to the majority of the other KIR genes ***(30)***.
6. In a population, where KIR genes have not been analyzed before, it is best to use SBT in case there are yet unknown alleles.
7. PCR-SSP typing requires plenty of DNA. If the amount of DNA is limited, whole genomic amplification can be successfully used for KIR typing ***(64)***.

References

1. Lanier, L. L. (2005) NK cell recognition. *Ann. Rev. Immunol.* **23,** 225–274.
2. Parham, P. (2005) MHC class I molecules and KIRs in human history, health and survival. *Nat. Rev. Immunol.* **5,** 201–214.
3. Karre, K., Ljunggren, H. G., Piontek, G. and Kiessling, R. (1986) Selective rejection of H-2-deficient lymphoma variants suggests alternative immune defence strategy. *Nature* **319,** 675–678.
4. Bauer, S., Groh, V., Wu, J., Steinle, A., Phillips, J. H., Lanier, L. L. and Spies, T. (1999) Activation of NK cells and T cells by NKG2D, a receptor for stress- inducible MICA. *Science* **285,** 727–729.
5. Borrego, F., Kabat, J., Kim, D. K., et al. (2002) Structure and function of major histocompatibility complex (MHC) class I specific receptors expressed on human natural killer (NK) cells. *Mol. Immunol.* **38,** 637–660.
6. Pende, D., Cantoni, C., Rivera, P., et al. (2001) Role of NKG2D in tumor cell lysis mediated by human NK cells: cooperation with natural cytotoxicity receptors and capability of recognizing tumors of nonepithelial origin. *Eur. J. Immunol.* **31,** 1076–1086.
7. Rajagopalan, S. and Long, E. O. (1997) The direct binding of a p58 killer cell inhibitory receptor to human histocompatibility leukocyte antigen (HLA)-Cw4 exhibits peptide selectivity. *J. Exp. Med.* **185,** 1523–1528.
8. Young, N. T., Rust, N. A., Dallman, M. J., Cerundolo, V., Morris, P. J., and Welsh, K.I. (1998) Independent contributions of HLA epitopes and killer inhibitory receptor expression to the functional alloreactive specificity of natural killer cells. *Hum. Immunol.* **59,** 700–712.
9. Katz, G., Gazit, R., Arnon, T. I., et al. (2004) MHC class I-independent recognition of NK-activating receptor KIR2DS4. *J. Immunol.* **173,** 1819–1825.
10. Wilson, M. J., Torkar, M., Haude, A., et al. (2000) Plasticity in the organization and sequences of human KIR/ILT gene families. *Proc. Natl. Acad. Sci. USA* **97,** 4778–4783.
11. Barten, R., Torkar, M., Haude, A., Trowsdale, J., and Wilson, M.J. (2001) Divergent and convergent evolution of NK-cell receptors. *Trends Immunol.* **22,** 52–57.
12. Trowsdale, J. (2001) Genetic and functional relationships between MHC and NK receptor genes. *Immunity* **15,** 363–374.

13. Shilling, H. G., Guethlein, L. A., Cheng, N. W., et al. (2002) Allelic polymorphism synergizes with variable gene content to individualize human KIR genotype. *J. Immunol.* **168,** 2307–2315.
14. Parham, P. (2004) Killer cell immunoglobulin-like receptor diversity: balancing signals in the natural killer cell response. *Immunol. Lett.* **92,** 11–13.
15. Uhrberg, M., Valiante, N. M., Shum, B. P., et al. (1997) Human diversity in killer cell inhibitory receptor genes. *Immunity* **7,** 753–763.
16. Toneva, M., Lepage, V., Lafay, G., et al. (2001) Genomic diversity of natural killer cell receptor genes in three populations. *Tissue Antigens* **57,** 358–362.
17. Norman, P. J., Stephens, H. A., Verity, D. H., Chandanayingyong, D., and Vaughan, R.W. (2001) Distribution of natural killer cell immunoglobulin-like receptor sequences in three ethnic groups. *Immunogenetics* **52,** 195–205.
18. Norman, P. J., Carrington, C. V., Byng, M., et al. (2002) Natural killer cell immunoglobulin-like receptor (KIR) locus profiles in African and South Asian populations. *Genes Immun.* **3,** 86–95.
19. Cook, M. A., Moss, P. A., and Briggs, D. C. (2003) The distribution of 13 killer-cell immunoglobulin-like receptor loci in UK blood donors from three ethnic groups. *Eur. J. Immunogenet.* **30,** 213–221.
20. Husain, Z., Alper, C. A., Yunis, E. J., and Dubey, D. P. (2002) Complex expression of natural killer receptor genes in single natural killer cells. *Immunology* **106,** 373–380.
21. Rajagopalan, S. and Long, E. O. (1999) A human histocompatibility leukocyte antigen (HLA)-G-specific receptor expressed on all natural killer cells. *J. Exp. Med.* **189,** 1093–1100.
22. Chan, H.-W., Kurago, Z. B., Stewart, C. A., et al. (2003) DNA methylation maintains allele-specific KIR gene expression in human natural killer cells. *J. Exp. Med.* **197,** 245–255.
23. Kikuchi-Maki, A., Yusa, S., Catina, T. L., and Campbell, K. S. (2003) KIR2DL4 is an IL-2-regulated NK cell receptor that exhibits limited expression in humans but triggers strong IFN-gamma production. *J. Immunol.* **171,** 3415–3425.
24. Marsh, S. G., Parham, P., Dupont, B., et al. (2003) Killer-cell immunoglobulin-like receptor (KIR) nomenclature report, 2002. *Hum. Immunol.* **64,** 648–654.
25. Hsu, K. C., Liu, X. R., Selvakumar, A., Mickelson, E., O'Reilly, R. J., and Dupont, B. (2002) Killer Ig-like receptor haplotype analysis by gene content: evidence for genomic diversity with a minimum of six basic framework haplotypes, each with multiple subsets. *J. Immunol.* **169,** 5118–5129.
26. Crum, K. A., Logue, S. E., Curran, M. D., and Middleton, D. (2000) Development of a PCR-SSOP approach capable of defining the natural killer cell inhibitory receptor (KIR) gene sequence repertoires. *Tissue Antigens* **56,** 313–326.
27. Maxwell, L. D., Wallace, A., Middleton, D., and Curran, M. D. (2002) A common KIR2DS4 deletion variant in the human that predicts a soluble KIR molecule analogous to the KIR1D molecule observed in the rhesus monkey. *Tissue Antigens* **60,** 254–258.

28. Hsu, K. C., Chida, S., Geraghty, D. E., and Dupont, B. (2002) The killer cell immunoglobulin-like receptor (KIR) genomic region: gene-order, haplotypes and allelic polymorphism. *Immunol. Rev.* **190,** 40–52.
29. Yawata, M., Yawata, N., Abi-Rached, L., and Parham, P. (2002) Variation within the human killer cell immunoglobulin-like receptor (KIR) gene family. *Crit. Rev. Immunol.* **22,** 463–482.
30. Garcia, C. A., Robinson, J., Guethlein, L. A., Parham, P., Madrigal, J. A., and Marsh, S. G. (2003) Human KIR sequences 2003. *Immunogenetics* **55,** 227–239.
31. Middleton, D. (2005) KIR allele and gene polymorphism group (KAG). *Mol. Immunol.* **42,** 455–457.
32. Robinson, J., Waller, M. J., Stoehr, P., and Marsh, S. G. (2005) IPD—the Immuno Polymorphism Database. *Nucleic Acids Res.* **33,** D523–D526.
33. Helmberg, W., Dunivin, R., and Feolo, M. (2004) The sequencing-based typing tool of dbMHC: typing highly polymorphic gene sequences. *Nucleic Acids Res.* **32,** W173–W175.
34. Valiante, N. M., Uhrberg, M., Shilling, H. G., et al. (1997) Functionally and structurally distinct NK cell receptor repertoires in the peripheral blood of two human donors. *Immunity* **7,** 739–751.
36. Gumperz, J. E., Valiante, N. M., Parham, P., Lanier, L. L., and Tyan, D. (1996) Heterogeneous phenotypes of expression of the NKB1 natural killer cell class I receptor among individuals of different human histocompatibility leukocyte antigens types appear genetically regulated, but not linked to major histocompatibililty complex haplotype. *J. Exp. Med.* **183,** 1817–1827.
37. Trompeter, H. I., Gomez-Lozano, N., Santourlidis, S., et al. (2005) Three structurally and functionally divergent kinds of promoters regulate expression of clonally distributed killer cell Ig-like receptors (KIR), of KIR2DL4, and of KIR3DL3. *J. Immunol.* **174,** 4135–4143.
38. Leung, W., Iyengar, R., Triplett, B., et al. (2005) Comparison of killer Ig-like receptor genotyping and phenotyping for selection of allogeneic blood stem cell donors. *J. Immunol.* **174,** 6540–6545.
39. Gardiner, C. M., Guethlein, L. A., Shilling, H. G., et al. (2001) Different NK cell surface phenotypes defined by the DX9 antibody are due to KIR3DL1 gene polymorphism. *J. Immunol.* **166,** 2992–3001.
40. Frohn, C., Schlenke, P., and Kirchner, H. (1997) The repertoire of HLA-Cw-specific NK cell receptors CD158 a/b (EB6 and GL183) in individuals with different HLA phenotypes. *Immunology* **92,** 567–570.
41. Tanamachi, D. M., Hanke, T., Takizawa, H., Jamieson, A. M., and Raulet, D. R. (2001) Expression of natural killer receptor alleles at different Ly49 loci occurs independently and is regulated by major histocompatibility complex class I molecules. *J. Exp. Med.* **193,** 307–315.
42. Huard, B. and Karlsson, L. (2000) KIR expression on self-reactive CD8+ T cells is controlled by T-cell receptor engagement. *Nature* **403,** 325–328.
43. Vales-Gomez, M., Erskine, R. A., Deacon, M. P., Strominger, J. L., and Reyburn,

H. T. (2001) The role of zinc in the binding of killer cell Ig-like receptors to class I MHC proteins. *Proc. Natl. Acad. Sci. USA* **98,** 1734–1739.

44. Khakoo, S. I., Thio, C. L., Martin, M. P., et al. (2004) HLA and NK cell inhibitory receptor genes in resolving hepatitis C virus infection. *Science* **305,** 872–874.
45. Carrington, M., Wang, S., Martin, M. P., et al. (2005) Hierarchy of resistance to cervical neoplasia mediated by combinations of killer immunoglobulin-like receptor and human leukocyte antigen loci. *J. Exp. Med.* **201,** 1069–1075.
46. Shilling, H. G., McQueen, K. L., Cheng, N. W., Shizuru, J. A., Negrin, R. S., and Parham, P. (2003) Reconstitution of NK cell receptor repertoire following HLA-matched hematopoietic cell transplantation. *Blood* **101,** 3730–3740.
47. Vitale, C., Chiossone, L., Morreale, G., et al. (2004) Analysis of the activating receptors and cytolytic function of human natural killer cells undergoing in vivo differentiation after allogeneic bone marrow transplantation. *Eur. J. Immunol.* **34,** 455–460.
48. Sasazuki, T., Juji, T., Morishima, Y., et al. (1998) Effect of matching of class I HLA alleles on clinical outcome after transplantation of hematopoietic stem cells from an unrelated donor. Japan Marrow Donor Program. *N. Engl. J. Med.* **339,** 1177–1185.
49. Ruggeri, L., Capanni, M., Urbani, E., et al. (2002) Effectiveness of donor natural killer cell alloreactivity in mismatched hematopoietic transplants. *Science* **295,** 2097–2100.
50. Velardi, A., Ruggeri, L., Moretta, A., and Moretta, L. (2002) NK cells: a lesson from mismatched hematopoietic transplantation. *Trends Immunol.* **23,** 438–444.
51. Parham, P. and McQueen, K. L. (2003) Alloreactive killer cells: hindrance and help for haematopoietic transplants. *Nat. Rev. Immunol.* **3,** 108–122.
52. Davies, S. M., Ruggieri, L., DeFor, T., et al. (2002) Evaluation of KIR ligand incompatibility in mismatched unrelated donor hematopoietic transplants. Killer immunoglobulin-like receptor. *Blood* **100,** 3825–3827.
53. Giebel, S., Locatelli, F., Lamparelli, T., et al. (2003) Survival advantage with KIR ligand incompatibility in hematopoietic stem cell transplantation from unrelated donors. *Blood* **102,** 814–819.
54. Beelen, D. W., Ottinger, H. D., Ferencik, S., et al. (2005) Genotypic inhibitory killer immunoglobulin-like receptor ligand incompatibility enhances the long-term antileukemic effect of unmodified allogeneic hematopoietic stem cell transplantation in patients with myeloid leukemias. *Blood* **105,** 2594–2600.
55. Hsu, K. C., Keever-Taylor, C. A., Wilton, A., et al. (2005) Improved outcome in HLA-identical sibling hematopoietic stem cell transplantation for acute myelogenous leukemia (AML) predicted by KIR and HLA genotypes. *Blood* **105,** 4878–4884.
56. Leung, W., Iyengar, R., Turner, V., et al. (2004) Determinants of antileukemia effects of allogeneic NK cells. *J. Immunol.* **172,** 644–650.
57. Vitale, M., Carlomagno, S., Falco, M., et al. (2004) Isolation of a novel KIR2DL3-specific mAb: comparative analysis of the surface distribution and function of KIR2DL2, KIR2DL3 and KIR2DS2. *Int. Immunol.* **16,** 1459–1466.

58. Gagne, K., Brizard, G., Gueglio, B., et al. (2002) Relevance of KIR gene polymorphisms in bone marrow transplantation outcome. *Hum. Immunol.* **63,** 271–280.
59. Gomez-Lozano, N. and Vilches, C. (2002) Genotyping of human killer-cell immunoglobulin-like receptor genes by polymerase chain reaction with sequence-specific primers: an update. *Tissue Antigens* **59,** 184–193.
60. Cook, M. A., Norman, P. J., Curran, M. D., et al. (2003) A Multi-Laboratory characterization of the KIR genotypes of 10th International Histocompatibility Workshop cell lines. *Hum. Immunol.* **64,** 567–571.
61. Meenagh, A., Williams, F., Sleator, C., Halfpenny, I. A., and Middleton, D. (2004) Investigation of killer cell immunoglobulin-like receptor gene diversity V. KIR3DL2. *Tissue Antigens* **64,** 226–234.
62. Witt, C. S., Martin, A., and Christiansen, F. T. (2000) Detection of KIR2DL4 alleles by sequencing and SSCP reveals a common allele with a shortened cytoplasmic tail. *Tissue Antigens* **56,** 248–257.
63. Sun, J. Y., Gaidulis, L., Miller, M. M., et al. (2004) Development of a multiplex PCR-SSP method for Killer-cell immunoglobulin-like receptor genotyping. *Tissue Antigens* **64,** 462–468.
64. Shao, W., Tang, J., Dorak, M. T., et al. (2004) Molecular typing of human leukocyte antigen and related polymorphisms following whole genome amplification. *Tissue Antigens* **64,** 286–292.
65. Sawicki, M. W., Dimasi, N., Natarajan, K., Wang, J., Margulies, D. H., and Mariuzza, R. A. (2001) Structural basis of MHC class I recognition by natural killer cell receptors. *Immunol. Rev.* **181,** 52–65.
66. Colucci, F., Di Santo, J. P., and Leibson, P. J. (2002) Natural killer cell activation in mice and men: different triggers for similar weapons? *Nat. Immunol.* **3,** 807–813.
67. Biassoni, R., Cantoni, C., Marras, D., et al. (2003) Human natural killer cell receptors: insights into their molecular function and structure. *J. Cell. Mol. Med.* **7,** 376–387.

11

Identification of Bone Marrow Derived Nonhematopoietic Cells by Double Labeling with Immunohistochemistry and *In Situ* Hybridization

Isinsu Kuzu and Meral Beksac

Summary

Stem cell migration/trafficking is a field of interest that is shared by pathologists, histologists, clinical transplantation teams, cardiologists, neurologists, and many other members of different disciplines. Until the findings of a successful combination of *in situ* methods, the origin of chimeric parenchymal cells was a dilemma. These double-labeling techniques have brought insight to our new concept of stem cell biology. It has been extremely helpful in the detection of the origin of terminally differentiated, including hematopoietic and nonhematopoietic, cells appearing following allogeneic stem cell transplantation. It has also become a standard approach for evaluation of repopulation following tissue injury in solid organ transplant patients or experimental models.

Although very useful, this technique has its advantages and pitfalls. It requires expertise in application and interpretation. Suitable selection of specific markers against parenchymal cells and preferably a cocktail of antibodies targeting infiltrating inflammatory cells are mandatory. One pitfall of this method is its restriction to sex-mismatched pairs. The spectrum of labels for X and Y chromosomes are suitable for combination. To prevent misinterpretation, the precautions needed are defined in this chapter.

Key Words: Immunohistochemistry; *in situ* hybridization; double labeling; chimerism; stem cell transplantation.

1. Introduction

The concept and our knowledge of stem cells has been almost completely altered during the last 5 yr. The exciting paper from Orlic et al. showing regeneration of infarcted myocardium with bone marrow stem cells in mice was the beginning of a new era and was followed by many experimental and clinical studies *(1)*. The initial reports implicating a stem cell plasticity/transdifferentiation created hope for treatment of disorders resulting from fatally malfunc-

From: *Methods in Molecular Medicine, vol. 134: Bone Marrow and Stem Cell Transplantation*
Edited by: M. Beksac © Humana Press Inc., Totowa, NJ

tioned organs or tissues. Unfortunately, these findings were later challenged by the reports supporting fusion between donor type adult hematopoietic stem cells and recipient's terminally differentiated cells of mesenchymal origin. Although the proof of diploid–diploid fusion by demonstration of tetraploid hepatocytes, purkinje neurons, or cardiomyocytes are beyond doubt, cell fusion is a very rare event, insufficient to explain the functional improvement observed ***(2–4)***. During the same time period, reports on demonstration of donor bone marrow-derived hepatic, neural, and other cells following allogeneic stem cell or organ transplantation in humans also continued to be published ***(5–9)***. The debate is still ongoing and the polarization in the scientific community has not come to a resolution yet. Another recent concept that has brought a new perspective into the field is that during early embryogenesis, when the migration and trafficking of stem cells are at their climax, stem cells anchor at various tissues. Damaged parenchymal cells can be replaced by lineage-specific tissue stem cells which are already located at sites called "niches." Liver, gastrointestinal mucosa, epidermis, endometrium, and cornea are some examples of the tissues with high capacity for self renewal. The lineage-specific tissue stem cells have restricted differentiation capacity compared to adult bone marrow multipotent progenitor cells. Today, there is accumulating evidence suggesting migration or circulation of stem cells in the postnatal period. The migration from the bone marrow to the peripheral blood after the infusion of growth factors, i.e., granulocyte colony-stimulating factor, has been known and used for therapeutic purposes for a long time. The reports showing engraftment of donor bone marrow-derived cells in nonhematopoietic tissues, i.e., liver, kidney, and brain, are the examples of evidence for traffic from the transplanted bone marrow to these tissues. The reverse has come to our attention very recently. It is also very likely that tissue-committed stem cells (TCSC) may migrate and colonize in the bone marrow, a very permissive environment, as well ***(10)***. This traffic helps to maintain a stem cell pool capable of cell renewal in all niches and serves many physiological repair purposes. The tissue damage may be a trigger for the trafficking of TCSCs into the circulation. The growth factors that mobilize stem cells from the marrow to the circulation have been found to mobilize bone marrow residing-TCSCs to the circulation as well ***(10)***. Under the light of these findings we may attribute the regenerative potential of bone marrow derived stem cells to these bone marrow-residing TCSCs and and conclude that it is not a transdifferentiation phenomenon. More evidence, other than fusion, against plasticity is the finding of very rare tissue-specific stem cells in the myocardium ***(11)***. The number of these stem cells may not be adequate enough for the regeneration needed following a myocardial infarct. However, the implantation of bone marrow mononuclear cells may establish an environment permissive for the cardiac stem cell proliferation and differentiation.

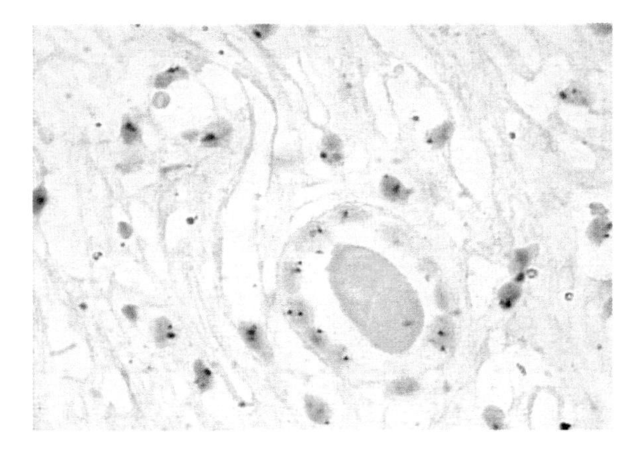

Fig. 1. Renal tubular epithelium with double X signals (red) of donor origin. The cells in the interstitial tissue contain one X (red) and one Y (green) signal demonstrating the recipient bone marrow origin. (CISH method alone for chromosomes X [red] and Y [green], DAP and Fast red). (Reproduced from **ref. *6***, with permission.)

Recently, reports from the Hannover group have demonstrated the donor origin of the Tc-labeled cells repopulating the infarct area by nuclear magnetic resonance imaging in patients ***(12)***. The origin of the donor bone marrow residing stem cells responsible for the regeneration has not been documented yet ***(13)***. In order to answer this and many other related questions, we need to analyze the posttransplant or postimplant tissues using markers specific to the donor with simultaneous phenotyping of all cells.

The finding that multipotent adult progenitor cells are not restricted to differentiation into hematopoietic cells has prompted investigators to apply this technique to the field of allogeneic stem cell or solid-organ transplantation. Disparities between donor–recipient pairs in terms of gender, molecular markers, etc. enables detection of chimerism in various cell types, including cells at different maturation steps, either hematopoietic or of other lineages. The most widely and easily applied technique is *in situ* hybridization (ISH). For simultaneous detection of various parameters, *in situ* methods including immunohistochemistry (IHC) and ISH which target proteins, DNA, or RNA, could be applied in combination. The origin and transdifferentiation status of the transplanted multipotent adult progenitor cells can be detected with the simultaneous application of these two techniques. For detection of donor type chimerism among non-hematopoietic cells in the parenchymal organ and tissues, or vice versa, a technique that enables detection of both genetic origin and differentiation is required. Chromogenic ISH (CISH) **(Fig. 1)** or fluorescence ISH (FISH) (**Figs. 2A** and **3A**) have all been used for this purpose. These techniques can be

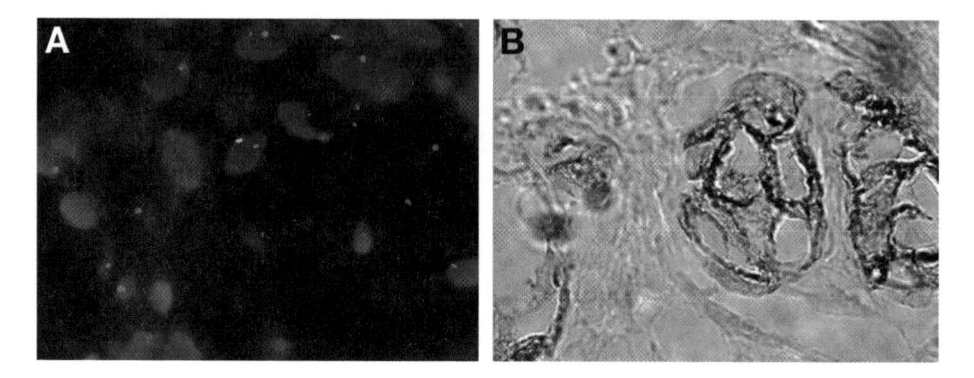

Fig. 2. **A** and **B** are from a colon mucosal biopsy sample of a female patient who had received bone marrow transplantation from a male donor. **(A)** A donor-originated epithelial cell with X (red) and Y (green) choromosomal signals within the crypt. The rest of the epithelial cells are recipient originated having double X signals (FISH method, dark field). **(B)** Cytokeratin expression (black-stained) of the donor originated cell in **A** is reflecting the epithelial phenotype (immunoperoxidase staining with diaminobenzidine chromogen, bright field).

successfully applied to frozen or paraffin-embedded tissue sections sequentially. The donor-derived inflammatory cell population in bone marrow-transplanted patients may infiltrate the parenchymal compartment. The main interpretation difficulty is caused by these overlapping bone marrow-derived inflammatory and host-derived parenchymal cells. Thus, labeling only parenchymal cells may cause overinterpretation of engraftment. In order to overcome this problem, double labeling with lineage-specific monoclonal antibodies and molecular probes has been used to define both the phenotype and genotype of the chimeric sample. The decision whether to use FISH or CISH is based on the availability of the relevant imaging system or need for additional filters. In situations requiring multiple surface or cytoplasmic targets for IHC, chromosomes can be stained by FISH ***(5,8,14)***. Although FISH is the most frequently used method, there are also various publications wherein CISH was used in the visualization of sex chromosomes ***(6,15–18)***. FISH alone has been routinely applied for more than 15 yr. This technique is standardized for labeling of mRNA, genes (oncogenes or the others), and chromosomes (whole or partial). It has been widely used for quantification of donor type chimerism in sex-mismatched stem cell transplants. Later, immunocytochemistry was combined with FISH on tissue sections and successful images were achieved (**Figs. 2** and **3**). Application of FISH to paraffin-embedded tissues to examine certain chromosomal abnormalities has become routine as well ***(19–24)***. Because the morphology of the

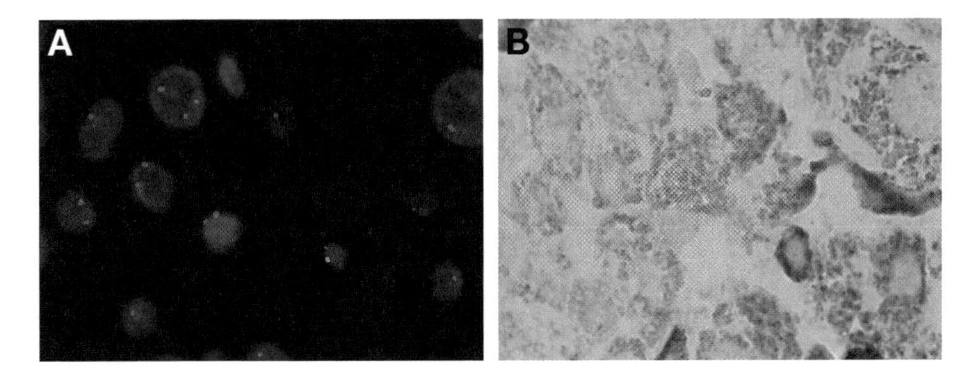

Fig. 3. The **A** and **B** are from a liver biopsy sample of a male patient who had received a liver transplant from a female living donor. **(A)** The donor-originated liver cells have double red signals of X chromosomes whereas recipient-originated bone marrow-derived inflammatory cell has one X (red) and one Y (green) chromosomal signals (FISH method, dark field). **(B)** The cell with male genotype (XY) in **A** is labeled (black) with the cocktail of antibodies against CD45, CD15, and CD68, revealing its bone marrow origin (Immuno peroxidase staining with DAB chromogen, bright field).

stained or unstained cells can be only roughly defined, an additional surface immunophenotyping step has become necessary.

The double-labeling method includes standard steps that are essential for tissue preparation. The pretreatment steps for ISH employ low pH, high temperature, and strong enzyme incubations. These conditions could damage the antigenic potential of the proteins used for cell labeling. For this reason, if double labeling is performed, pretreatment steps should follow IHC. If ISH is performed alone, the pretreatment procedure must follow **deparafinization**. The technique can be altered to the type of tissue and cells of interest. There are now biotin- and digoxygenin-labeled probes commercially available for CISH targeting of various genes including the X and Y chromosomes.

2. Materials

2.1. Specific Markers Against Cells and Tissues to be Selected

Specific markers for the parenchymal cell component can be extended depending on the tissues of interest. We could give a short list of tissues with their reliable markers. This list can be extended for different research purposes.

1. Liver: hepatocyte antibody.
2. Gastrointestinal epithelium: cytokeratin.
3. Neuroendocrine cells: synaptophysine.
4. Thyroid: thyroglobulin and cytokeratin.

5. Kidney: cytokeratin for proxymal tubular epithelium.
6. Glial cells: GFAP.
7. Neurons: neurofilament protein.
8. Bone marrow-derived inflammatory cells: mixture of several antibodies against different types of inflammatory cells that have similar antigen retrieval conditions can be applied on one section. Labeling of several different circulating or homing bone marrow-derived inflammatory cells can be easily visualized by this way. A mixture of anti-CD45, -CD68, and -CD15 can be applied for labeling of lymphocytes, monocytes- histiocytes, and leukocytes. All of the antibodies in the combination must be suitable for a common antigen-retrieval procedure.
9. Endothelium: factor VIII RA.
10. Striated muscle and cardiomyocyte myoglobin.

2.2. Tissue Sectioning and Deparaffinization

1. Microtome and blades for paraffin sectioning.
2. Positively charged slides (Superfrost plus, Menzel).
3. Xylene.
4. Ethanol (graded ethanol solutions: %75, %85, %100).
5. Incubator (essential temperatures are 37°C and 75°C).

2.3. Materials for IHC

1. % 3 H_2O_2 (v/v) solution
2. Antigen retrieval solutions: the most frequently used alternatives are heating in salt solutions, such as sodium citrate.

   Sodium citrate stock solution:

   Solution 1: 10.5 g citric acid ($C_2H_2O_2 \cdot H_2O$) in 500 mL dH_2O, pH 7.3.

   Solution 2: 29.41 g trisodium citrate ($C_6H_5Na_3O_7 \cdot 2H_2O$) in 1 L dH_2O, pH 7.3.

   Working solution: 3 mL of solution 1, 147 mL of solution 2 must be mixed in 1350 mL dH_2O.

   EDTA, pH 8.0 stock solution:

   Solution 1: 3.6 g disodium salt EDTA ($C_{10}H_{14}N_2Na_2O_8 - 2H_2O$) in 100 mL dH_2O, pH 8.0.

   Solution 2: 3.8 g tetrasodium salt EDTA ($C_{10}H_{12}N_2O_8Na_4 - 2H_2O$) in 100 mL dH_2O, pH 8.0.

   Working solution: 4.9 mL of solution 1, 5.1 mL of solution 2 must be mixed in 990 mL dH_2O.
3. Enzymes (Protease or trypsin).
4. Pressure cooker.
5. Primary antibodies: monoclonal or polyclonal (*see* **Subheading 2.1.**).
6. Secondary antibodies (anti-rabbit, anti-mouse, etc., depending on the primary antibody type) (*see* **Note 1**).

7. Phosphate-buffered saline (PBS) solution: 7.2 g NaCl, 0.43g NaH_2PO_4 $2H_2O$, 1.48 g Na_2 HPO_4 12 H_2O, in 1 L dH_2O and pH 7.6.
8. Diaminobenzidine (DAB).
9. Humid chamber.

2.4. FISH

1. HCl 0.2 *N* (Merck % 37 HCl):
   a. 1 *N* HCl stock solution: dilute 37 % HCl (w/v) 1:12 with dH_2O. This solution is stored at room temperature.
   b. 0.2 *N* HCl working solution: dilute 1 *N* HCl stock solution 1:5 with dH_2O. This solution can be used for 5–6 times when stored at room temperature.
2. 20X SSC stock solution: 3 *M* sodium chloride (NaCl), 0.3 *M* sodium citrate · $2H_2O$. Stock solution: dissolve 175.3 g NaCl and 88,2 g sodium citrate · $2H_2O$ in 800 mL dH_2O. Adjust pH to 7.0 with 1 *N* HCl. The final volume should be completed to 1 L by adding dH_2O. The stock solution can be stored at room temperature.
3. 2X SSC working solution: dilute 20X SSC 1:10 with dH_2O. This solution should be discharged at the end of the procedure.
4. 2X SSC working solution with Tween 20: 2X SSC with 0.05% Tween 20 (v/v).
5. Tris-EDTA-NaCl (TEN) buffer: 0.05 *M* Tris-HCl, pH 7.8, 0.01 *M* EDTA, 0.01 *M* NaCl. This solution can be stored at 4°C for a few weeks.
6. Formamide denaturation solution (70% formamide in 2X SSC). For 100 mL solution: 70 mL formamide + 10 mL 20X SSC+ 14 mL dH_2O. Adjust pH to 7.0 by 1 *N* HCl. Add dH_2O and complete the volume to 100 mL. This solution can be used five to six times when it is stored at 4°C.
7. Formamide washing solution (50 % formamide in 2X SSC). For 150 mL solution: 75 mL formamide + 15 mL 20X SSC + 40 mL dH_2O. Adjust pH to 7.0 by 1 *N* HCl. Add dH_2O and complete the volume to 150 mL. This solution can be used five to six times when it is stored at 4°C.
8. 4% Formaldehyde solution in PBS: dilute 40% formaldehyde (w/v) solution 1:10 in PBS. This solution can be used five to six times when it is stored at room temperature.
9. Proteinase K solution: includes 0.05 mg/mL proteinase K in TEN buffer.
10. Probes for X and Y chromosomes. Direct fluorescent-labeled Vysis Cep X, Y probe cocktails (Abbott Vysis, IL) have been used in our studies. Other probe alternatives for FISH are direct-labeled Chr X,Y probes of FISH by Q-BIOgene, CA. There is a commercially available product of Zymed-Invitrogen, CA, named Spot-Light Chr X/Y probe cocktail, which contains digoxgenin- and biotin-labeled probes. This product is suitable for CISH and the detection kit can be used following the probe incubation if this probe cocktail is performed.
11. DAPI is preferred for nuclear counterstaining for FISH.
12. Water bath suitable to reach boiling temperatures.
13. Probe Incubator (1-VP 2000 and HyBrite Vysis, Abott, 2-Omni Slide thermal cycler).

14. Glass coverslips.
15. Rubber cement.
16. Metal containers (these are better than the glass containers and help the solutions to reach high temperatures in the water bath).
17. Metal slide racks.

3. Methods

3.1. Paraffin Sections

In routine practice, 4- to 5-μm thick paraffin sections are preferred. This thickness may affect the interpretation by increasing the frequency of partial nuclei.

The range of nuclear losses depends on the size of the nuclei. The effect of sections on interpretation will be discussed later. The efficiency of slide adhesivity is also important. For this reason, the sections must be placed on positively charged slides in order to avoid losing the tissue during the procedure's several temperature and enzyme challenges.

3.2. Deparafinization

1. Incubate paraffin-embedded tissue sections (4–5 μm in thickness) at 37°C overnight. This procedure is useful for drying and making the tissue section to adhere to the slide.
2. Incubate paraffin sections at 70°C for 1–2 h.
3. Immerse slides in three changes of %100 xylene for 6 min each.
4. Rehydrate slides in graded ethanol solutions in three dilutions of 100%, 85%, and 75% (v/v) for 5 min each.
5. Wash the slides in dH_2O for at least 5 min before starting the IHC steps.

3.3. IHC

1. Treat slides with 3% H_2O_2 (v/v) for 5 min in order to block endogenous peroxidase.
2. Wash slides in two changes of PBS.
3. Antigen retrieval (this step is optional [*see* **Note 2**]) using pressure cooker: boil the solution (either sodium citrate or EDTA) and place the slide racks with the slides into the cooker. The slides should be completely immersed in the liquid. Close the lid and boil the slides under pressure for 90 to 120 s. At the end of the incubation period, immediately cool the cooker under tap water before opening the lid. Remove the slide racks and place in dH_2O.
4. Rinse slides in PBS for 5 min.
5. Wipe the liquid surrounding the section and circle the section with a delimitating pen in order to avoid the diffusion of the antibodies to the whole surface of the slide and keep them on the section.
6. Apply the primary antibody in a recommended dilution, period, and temperature (*see* **Subheading 2.1.**).
7. Rinse slides in two changes of PBS for 5 min each following the incubation period.

Table 1
Durations of Pretreatment Steps According to the Different Tissues *(19)*

Tissues	HCL (0.2 N) Incubation (min)	2X SSC at 80°C (min)	Proteinase K (0.05 mg/mL) at 37°C (min)
• Skin • Intestine • Lung • Thyroid	15 (± 5)	15 (± 5)	5–10
• Spleen • Lymph node • Adrenal	20	20	10
• Liver • Kidney • Pancreas • Heart • Breast [a] • Prostate [a]	25 (± 5)	25 (± 5)	10–15

[a] The periods for breast and prostate tissues are obtained by our laboratory experience.

8. Apply secondary antibody for the required period and temperature (*see* **Note 1** for the alternatives).
9. Rinse slides twice with PBS for 5 min each.
10. Apply enzyme substrate DAB for required time (about 10–20 min depending on the product manual) at room temperature (*see* **Note 4**).
11. Rinse the slides twice with PBS for 5 min each (*see* **Note 5**).
12. Rinse the slides in dH_2O.

3.4. FISH

3.4.1. Tissue Pretreatment Protocol

The pretreatment procedure is important especially when paraffin sections of fixed tissues are used. The parenchymal cells of every tissue are placed in a stroma containing extra cellular matrix. The cellular membrane and the amount of cytoplasm could alter the efficiency of the hybridization. The probes must pass through all of these structures, such as extra cellular matrix, cellular membrane, cytoplasm, and nuclear membrane, and finally reach the chromosomes. For this reason, different incubation periods for different tissues are recommended for the key steps of the pretreatment procedure (**Table 1**). For a successful result, the tissue and cell structure must be optimally damaged. The aim is to make enough holes for the probes to reach the chromosomes during the incubation without damaging the morphology. The fixatives can also affect the success of the hybridization.

In our experience, the protocol most effective on several different tissues is that described by Johnson and colleagues ***(19)***. In order to help the denaturation of the cellular DNA and increase the hybridization success, we modified this method by adding a formamide denaturation step described earlier ***(23,24)***. The following part contains pretreatment steps first described by Johnson and colleagues, including our modification ***(23,24)***. The pretreatment procedure explained here must follow IHC, but can also be used prior to probe hybridization on paraffin-embedded tissue sections to examine cellular DNA for purposes excluding IHC combination.

1. Incubate the slides in 0.2 *M* HCl at room temperature. Different durations of the incubation period are recommended for different tissues ***(19)*** (*see* **Table 1**).
2. Rinse the slides in dH_2O at room temperature for 2 min by dipping several times.
3. Rinse slides in 2X SSC with Tween 20 for 5 min at room temperature.
4. Incubate slides with 2X SSC at 80°C for recommended time (*see* **Table 1**). Containers with 2X SSC solution must be heated previously in the water bath and the slides must be immersed at 80°C.
5. Rinse the slides in dH_2O at room temperature for 2 min by dipping several times.
6. Rinse slides in 2X SSC with Tween 20 for 5 min at room temperature.
7. Incubate slides with proteinase K in TEN buffer at 37°C in a humid chamber (*see* **Table 1** for the period of treatment).
8. Rinse the slides in dH_2O at room temperature for 2 min by dipping several times.
9. Rinse slides in 2X SSC with Tween 20 for 5 min at room temperature.
10. Incubate slides with 4% formaldehyde solution for 10 min at room temperature.
11. Rinse slides in 2X SSC with Tween 20 for 5 min at room temperature.
12. Treat slides with formamide denaturation solution at 74°C for 5 min
13. Rinse slides in 2X SSC with Tween 20 for 5 min at room temperature.
14. Dehydrate the slides in three changes of graded ethanol solutions; 70%, 85%, and 100% (v/v) for 2 min each.
15. Leave the slides to dry at room temperature for at least 10 min (*see* **Note 6** for storing conditions for the stained sections).

3.4.2. Probe Incubation

1. Apply probe mixtures of X-Y chromosomes (Cep X-Y probes, Vysis, IL are used in our studies) to the sections (*see* **Subheading 2.4., item 10** for probe alternatives and **Note 7** for determining the amount of probe solution).
2. Place a glass coverslip on the section and seal the edges with rubber cement. Leave the cement to dry before the next step.
3. Place the slides on the hot plate or probe incubator. The probes mentioned previously must be heat denatured at 90°C for 10 min, and incubated at 42°C overnight. (These conditions and the time periods are suitable for the mentioned Vysis

Cep X,Y probes. Other probes can be incubated in different conditions as recommended in their data sheets.)

3.4.3. Hybridization Washings

1. Incubate the slides in 2X SSC at 42°C for 10–20 min to hydrate the rubber cement.
2. Remove the rubber cement gently, leaving the coverslip on the section. Tissue sections can be lost if the coverslip changes its position or is pulled away from the surface by force during this procedure. Replace the slides in the same solution and allow the coverslip to leave the slide by dipping the slides several times in the buffer solution.
3. Following the complete removal of the coverslip, wash the slides in formamide washing solution for 10 min at 42°C.
4. Wash the slides in 2X SSC at 42°C for 5 min.
5. Mount the slides with an antifade containing DAPI solution.
6. Seal the edges of the coverslip either with transparent nail polish or rubber cement before the examination (*see* **Note 8**).
7. For examination of the slides, a fluorescence microscope with suitable filters is necessary. High-quality images can be obtained when and image analysis system is attached.

3.5. Interpretation

Interpretation of the results is the last step of double labeling. We would like to discuss the advantages and the pitfalls of the different ISH methods for this purpose.

Either FISH or CISH probes for Chromosomes X, Y can be used for detection of chimerism on paraffin embedded tissue sections. The advantage of CISH compared to FISH is its suitability to allow for multiple examinations of slides under the light microscope (**Fig. 1**). The pitfall could be the time-consuming steps for visualization with secondary enzyme-conjugated systems during the staining procedure. The steps of the FISH procedure are shorter than those of CISH and the contrast of the signal is better on the dark field. FISH examination must be done under the florescence microscope on a dark field, which may cause difficulty in orientation to the cellular architecture. Quick fading of the fluorescence signals, which disables several examinations, is another pitfall. To summarize, the two ISH methods have their advantages and pitfalls when they are applied on paraffin-embedded tissue sections. Neither of them, when used alone, is fully sufficient for obtaining reliable information about chimerism in parenchymal cells and tissues.

When immunoenzymatic methods for IHC are combined with FISH, simultaneous examination of either the phenotype (which describes the cellular origin) or the genotype (describes the chimerism status) of each cell can be demonstrated reliably. During the examination, results of these two methods can be interpreted on a dark field for FISH signals and on a bright field for IHC staining, without changing the area. Images of the area can be captured with dark field microscopy with suitable filters and bright field microscopy for demonstration of the results (*see* **Fig. 2A,B**). By using alternative chromogens (fast red, fast blue, etc.), combination of enzyme-conjugated IHC and CISH can also give reliable results. The examples of other combination methods are present in the literature ***(8,14–18)***.

The interpretation of the cellular origin of the chimeric cells is the most important point for proving the engraftment of donor type stem cells in extramedullary sites in sex-mismatched bone marrow-transplanted individuals. The reverse has been demonstrated with recipient bone marrow-originated cell repopulation in graft parenchyma following solid organ transplantation ***(5,6)***. The chromosomal signals from superposing bone marrow-derived inflammatory cells may cause overdiagnoses of parenchymal cell repopulation. Staining the sequential sections by specific markers for parenchymal cells may not be enough, and a cocktail of inflammatory cell markers may be preferable in order to avoid misinterpretation due to overlapping bone marrow-originated inflammatory cells (**Figs. 2** and **3**).

Another factor that could influence the interpretation may result from paraffin sectioning. Although confocal microscopy is the best method for three-dimensional examination of cells and their nuclei, the advanced imaging systems may not be possible in every center. Neither of these sophisticated systems is suitable for routine pathology examinations. In order to decrease the likelihood of misinterpretation due to paraffin sectioning in different FISH applications, the investigators suggest counting only the cells that express the expected number of chromosomal signals ***(21,22)***. The image analysis systems suitable for routine FISH interpretation enable us to examine the sections on seven to eight sequential levels and allow examination of the signal on different levels in the range of the section. The misinterpretation due to losses by sectioning may cause underestimation of repopulation rate rather than overestimation. The interpretation of chimerism in extranodal sites detected by combination of *in situ* methods will be more reliable when the previously mentioned points are considered.

4. Notes

1. There are several alternative commercially available secondary visualization systems in the market. Even automated immunostaining systems (such as Ventana-Benchmark, or others) can be used for IHC. The focus here should be on avoiding the nuclear counterstaining, which could negatively affect the further ISH steps and interpretation. The secondary antibodies can be directly conjugated by HRP or biotin. If direct HRP conjugation of the secondary antibody is performed, the fol-

lowing step will be the substrate application. If biotin labeling of the secondary antibody is performed, a third step employing anti-biotin avidin complex must be performed before the substrate. Between each step, washing with PBS is required.
2. Antigen retrieval methods depend on the technical characteristics of the primary antibodies. A microwave oven or pressure cooker can be used for achieving high temperatures during the retrieval process. In our laboratory, a pressure cooker is preferred in the case of the requisite antigen retrieval with Citrate and EDTA solutions.
3. In order to avoid nonspecific binding of tissue proteins with the antibodies, blocking solutions containing normal serum can be applied, if necessary, before this step. Primary antibody must be applied after dripping the blocking reagent, without rinsing in between.
4. Because dehydration will be applied at the following step, chromogens resistant to ethanol must be selected. DAB is the ideal chromogen for this purpose. Other alternative AEC must be avoided, as it will vanish during the ethanol dehydration prior to probe incubation.
5. Nuclear counterstaining by hematoxylin must be avoided following this step.
6. After the slides are dry, they can be wrapped in aluminum foil or stored in slide boxes, which keep them away from dust. In these conditions, the slides can be stored for several weeks at room temperature before probe application.
7. The amount of probe solution is related to the size of the tissue section and the glass coverslip. If the section area is covered by 22 × 22 mm coverslips, at least 15–20 µL probe solution, depending on the tissue section size, will be needed for the procedure. When the area is covered by 15 × 15 mm coverslips, depending on the size of the tissue sections, 5–10 µL probe mix will be enough. This and the following steps must be performed in a dark room when directly fluorescence-labeled probes are used.
8. Sealing the edges of the coverslip is useful for avoiding evaporation and loss of tissue due to movement of the coverslip during the examination.

References

1. Orlic, D., Kajstura, J., Chimenti, S., et al. (2001) Bone marrow stem cells regenerate infarcted myocardium. *Nature* **410,** 701–705.
2. Terada, N., Hamazaki, T., Oka, M., et al. (2002) Bone marrow cells adopt the phenotype of other cells by spontaneous cell fusion. *Nature* **416,** 542–545.
3. Alvarez-Dolado, M., Pardal, R., Garcia-Verdugo, J. M., et al. (2003) Fusion of bone-marrow derived cells with purkinje neurons, cardiomyocytes and hepatocytes. *Nature* **425,** 968–973.
4. Nygren, J. M., Jovinge, S., Breitbach, M., et al. (2004) Bone marrow derived hematopoietic cells generate cardiomyocytes at a low frequency through cell fusion, but not transdifferentiation. *Nat. Med.* **10,** 494–501.
5. Idilman, R., Erden, E., Kuzu, I., et al. (2004) Recipient derived hepatocytes in sex mismatched liver allografts after liver transplantation: early versus late transplant biopsies. *Transplantation* **78,** 1647–1652.

6. Mete, T., Ensari, A., Sengul, S., Keven K., and Erbay B.(2005) The presence of recipient-derived renal cells in damaged allograft kidneys. *Nephrology Dialysis Transplant.* **20 (Suppl 5),** 361.
7. Cogle, C. R., Yachnis, A. T., Laywell, E. D., et al. (2004) Bone marrow transdifferentiation in brain after transplantation. *Lancet* **363,** 1432–1437.
8. Idilman, R., Kuzu, I., Erden, E., et al. (2006) Evaluation of the effect of transplantatıon related factors and tissue ınjury on donor-derived hepatocyte and gastrointestinal epithelial cell repopulation following hematopoietic cell transplantation. *Bone Marrow Transplant.* **37,** 199–206.
9. Matsumoto, T., Okamoto, R., Yajima, T., et al.(2005) Increase of bone marrow-derived secretory lineage epithelial cells during regeneration in the human intestine. *Gastroenterology* **128,** 1852–1867.
10. Kucia, M., Reca, R., Jala, V. R., Dawn, B., Ratajczak, J., Ratajczak, M. Z. (2005) Bone marrow as a home of heterogenous populations of nonhematopoietic stem cells. *Leukemia* **19,** 1118–1127.
11. Urbanek, K., Torella, D., Sheikh, F., et al. (2005) Myocardial regeneration by activation of multipotent cardiac stem cells in ischemic heart failure. *Proc. Natl. Acad. Sci. USA* **102,** 8692–8697.
12. Küstermann, E., Roell, W., Breitbach, M., et al. (2005) Stem cell implantation in ischemic mouse heart: a high-resolution magnetic resonance imaging investigation. *NMR Biomed.* **18,** 362–370.
13. Blau, H. M., Brazelton, T. R., and Weimann, J. M.(2001) The evolving concept of a stem cell: entity or function. *Cell* **105,** 829–841.
14. Brittan, M., Braun, K. M, Reynolds, L. E, et al. (2005) Bone marrow cells engraft within the epidermis and proliferate in vivo with no evidence of cell fusion**.** *J. Pathol.* **205,** 1–13.
15. Poulsom, R., Alison, M. R., Forbes, S. J., and Wright, N. A. (2002) Adult stem cell plasticity. *J. Pathol.* **197,** 441–456.
16. Direkze, N. C., Hodivala-Dilke, K., Jeffery, R., et al. (2004) Bone marrow contribution to tumor-associated myofibroblasts and fibroblasts. *C ancer Res.* **64,** 8492–8495.
17. Poulsom, R., Alison, M. R., Cook, T., et al. (2003) Bone marrow stem cells contribute to healing of the kidney. *Am. Soc. Nephrol***. 14(Suppl 1),** S48–54.
18. Direkze, N. C., Forbes, S. J., Brittan, M., et al (2003) Multiple organ engraftment by bone-marrow-derived myofibroblasts and fibroblasts in bone-marrow-transplanted mice. *Stem Cells* **21,** 514–520.
19. Johnson, K. L., Zhen, D. K., and Bianchi, D. W. (2000) The use of fluorescence in-situ hybridization (FISH) on paraffin-embedded tissue sections for study of microchimerism. *BioTechniques* **29,** 1220–1224.
20. Yong-Jie, L., Birdsall, S., Summersgill, B., et al. (1999) Dual Colour Fluorescence in situ Hybridization to parafin-embedded samples to deduce the presence of the der(X) t(X;18)(p 11.2;q11.2) and involvement of either the SSX1 or SSX2 gene: A Diagnostic and prognostic aid for synovial sarcoma. *J. Pathol.* **187,** 490–496.

21. Haralambieva, E., Kleiverda, K., Mason, D. Y., Schuuring, E., and Kluin, P. M. (2002) Detection of three common translocation breakpoints in non-Hodgkin's lymphomas by fluorescence in situ hybridization on routine paraffin-embedded tissue sections. *J. Pathol.* **198,** 163–170.
22. Haralambieva, E., Alison, H. B, Bastard, C., et al. (2003) Detection by the fluorescence in situ hybridization technique of MYC translocations in parafin-embedded iymphoma biopsy samples. *Br. J. Haem.* **121,** 49–56.
23. Speel, E. J. M. (1999). Detection and amplification systems for sensitive, multiple-target DNA and RNA in situ hybridization: looking inside cells with a spectrum of colors. *Histochem. Cell Biol.* **112,** 89–113.
24. Matsuta, M. and Matsuta, M (2000) In situ hybridization for DNA: fluorescent probe, in *Molecular Histochemical Techniques* (Koji., T., ed.). Springer, Germany, pp. 66–99.

12

Molecular Methods Used for Detection of Minimal Residual Disease Following Hematopoietic Stem Cell Transplantation in Myeloid Disorders

Ahmet H. Elmaagacli

Summary

Monitoring of minimal residual disease (MRD) in patients with acute or chronic myeloid disorders is routinely performed after allogeneic or autologous transplantation. The detection of MRD helps to identify patients who are at high risk for leukemic relapse after transplantation. The most commonly used techniques for MRD detection are qualitative and quantitative PCR methods, fluorescence *in situ* hybridization (FISH), florescence-activated cell sorting (FACS), and cytogenetic analysis, which are often performed complementary in order to assess more precisely MRD. Here, we describe the most used sensitive real-time reverse-transcription (RT)-PCR methods for chronic and acute myeloid disorders. Besides protocols for real-time RT-PCR and multiplex RT-PCR procedures for the most common fusion-gene transcripts in acute and chronic myeloid disorders, methods for detection of disease-specific genetic mutated alterations as FLT3 gene-length mutations, and aberrantly expressed genes as WT1 gene transcripts, are described in detail for daily use.

Key Words: MRD; transplant; real-time RT-PCR; myeloid disorders; fusion-transcripts; FLT3 length mutations; WT1.

1. Introduction

The invention of PCR and its implementation in hematology was a milestone in the detection of minimal residual disease (MRD). Nevertheless, qualitative detection of MRD is often not sufficient to assess precisely the individual risk for clinical relapse in patients with acute or chronic myeloid disorders after transplantation because not every qualitative detection of MRD is followed by a clinical relapse *(1–4)*. Indeed, quantitative methods are able to predict a clinical relapse more reliable by monitoring the kinetics of increasing MRD than qualitative methods. The requirement for a good MRD method is not only limited to

From: *Methods in Molecular Medicine, vol. 134: Bone Marrow and Stem Cell Transplantation*
Edited by: M. Beksac © Humana Press Inc., Totowa, NJ

Table 1
Most Common Minimal Residual Disease Markers in Chronic and Myeloid Leukemia

Disease	Type of marker	Chromosomal changes	Molecular changes
CML,AML,ALL	chromosomal aberration	translocation (9;22)	BCR-ABL (fusiongene)
AML FAB M2	chromosomal aberration	translocation (8;21)	AML1-ETO (fusiongene)
AML FAB M3	chromosomal aberration	translocation(15;17)	PML-RARA (fusiongene)
AML FAB M4	chromosomal aberration	inversion (16)	CBFβ-MYH11 (fusiongene)
AML FAB M5	chromosomal aberration	translocation involving chromosome 11q23	MLL
AML	length mutation of FLT3 Gene		FLT3-LM
AML, ALL, CML	aberrantly over-expressed Wilms tumor 1gene		WT-1

CML, chronic myeloid leukemia; AML, acute myeloid leukemia; ALL, acute lymphoblastic leukemia; MLL, mixed-lineage leukemia.

their ability to quantify MRD, but also to a need for a high sensitivity in order to predict a relapse as early as possible. These both important criteria for MRD detection are fulfilled by the real-time reverse-transcription (RT)-PCR ***(5)***. Nevertheless, qualitative PCR methods and other techniques as fluorescence *in situ* hybridization (FISH), florescence-activated cell sorting (FACS), and cytogenetic analysis are further pillars of MRD detection in hematology ***(6–8)***. However, they should not be regarded as competitive methods to PCR, but moreover as complementing techniques in the detection and assessment of MRD. Targets for MRD detection by PCR are fusion-gene transcripts, breakpoint regions around molecular rearrangements, and aberrantly expressed genes. **Table 1** shows the frequently expressed fusion genes in acute and chronic myeloid leukemia.

Real-time RT-PCR is the most sensitive method for the detection of MRD with a detection level up to $1{:}10^{-5}$ ***(5)***. Real-time RT-PCR measures the mRNA expression level of a target gene relative to one or more other reference genes. The used RNA per sample or cell may vary greatly in each real-time RT-PCR as a result of the quality of the sample, the elapsed time from obtaining the sample to RNA preparation, and many other factors. Also, the cDNA yield may vary additionally as a result of variation in enzyme activities or inhibitors in the sample. To normalize these quality variations, each target mRNA should be quantified in relation to the expression of so-called housekeeping genes which are expressed at similar levels in all different cell types. Commonly, two different real-time RT-PCR methods are performed based on the real-time RT-PCR method using hybridization probes. One real-time PCR method is based on the Taqman™ technology using Taqman probes and the other one on the Lightcycler® method using hybridization probes.

The Taqman probe has a 5′ fluorescent reporter dye, which is quenched by a 3′ quencher dye through fluorescence resonance energy transfer (FRET). After hybridization to the template, the fluorigenic Taqman probes will be hydrolyzed during PCR by the 5′ secondary structure dependent nuclease activity of the Taq DNA polymerase ***(10)***. After hydrolysis, the release of the reporter signal causes an increase in fluorescence intensity that is proportional to the accumulation of the PCR product. The Lightcycler method uses two hybridization probes, which hybridizes at the template closely together with a distance of only 1 to 5 bases. The first probe is fluorescent labeled at the 3′ and the second probe at the 5′site. Only after specific hybridization of both probes together to the template, a fluorescent signal is released which is proportional to the accumulation of the PCR product like the Taqman probe.

Both systems then generate a real-time amplification plot based upon the normalized (background) fluorescence signal. Subsequently a threshold cycle is determined, i.e., the fractional cycle number at which the amount of amplified target reaches a fixed threshold. This threshold is defined in general as three to ten times the standard deviation of the background fluorescent signal. When plotting the fluorescent signal vs the cycle number, the higher the initial copy number is, the earlier the signal appeared out of the defined threshold line. The "crossing points" are used to create a standard curve by plotting known concentration of serial dilutions of a control mRNA vs the "crossing points." The target mRNA expression in patients samples can be related to a standard curve as the number of nanograms of a control RNA (plasmid or leukemic cell line with the specific target fusion gene) with the same level of expression or in numbers of the copies of the target gene. In the following, the most common used qualitative and quantitative PCR methods for acute and chronic myeloid disorders are described in detail.

2. Materials

2.1. Equipment

1. Lightcycler device (Roche Diagnostics, Mannheim, Germany) (*see* **Note 1**).
2. ABI Prism Genetic Analyzer 310 including Genemapper v.3.10 software (Applied Biosystems, Darmstadt, Germany).

3. Capillars for Lightcycler device (Roche Diagnostics).
4. Cooling centrifuge for 2 mL Eppendorf tubes (12000 g) (Eppendorf Hamburg).
5. Agarose gel electrophoresis apparatus (Bio-Rad).
6. Ultraviolet light (Hofer Scientific Instruments, San Francisco, CA).
7. Spectrometer Lambda Bio for RNA/DNA measurement (Applied Biosystems).
8. MagNAPur™ Compact (Roche Diagnostics, Mannheim, Germany).

2.2. Reagents

2.2.1. RNA Isolation

1. RNeasy® Mini Kit (Qiagen, Hilden, Germany).
2. Erythrocyte-lysis buffer (Qiagen, Hilden).
3. RLT-Buffer (Qiagen, Hilden).

2.2.2. DNA Isolation

MagNA Pure compact nucleic acid isolation kit I (Roche Diagnostics).

2.2.3. Reverse Transcription and cDNA-Synthesis

1. Ultrapure desoxynucleosidetriphosphate (dNTP)–Set 100 m*M* (Pharmacia, Uppsala, Sweden). Add 20 μL of each of the four different dNTP:

dATP, dCTP, dGTP, and dTTP	80.0 μL
ad bdH_2O	1520 μL

2. bdH_2O (Aqua ad injectabilia Braun, Melsungen, Germany).
3. M-MLV Reverse Transcriptase 200 U/μL (Gibco Life Technologies, Eggenstein).
4. rRNAsin-Ribonuclease-Inhibitor 40 U/μL (Promega AG, Heidelberg).
5. Pd (N)6, random Hexamere 10 p*M*/μL (Pharmacia, Uppsala, Schweden).

2.2.4. RT-PCR and PCR

1. Oligonucleotide primers (Invitrogen).
2. Ampli Taq-DNA-Polymerase 5 U/ μL (Applied Biosystems).
3. Ampli Taq-Gold-Polymerase 5 U/ μL (Applied Biosystems).
4. PCR-buffer includes $MgCl_2$ 25 m*M* solution (Applied Biosystems).
5. SeaKem® LE agarose (FMC Bioproducts, Rochland, ME).
6. Ethidium bromide (Sigma, Muenich, Germany).
7. bdH_2O (Aqua ad in jectabilia Braun, Melsungen, Germany).

2.2.5. Real-Time PCR

1. Taqman and hybridization probes (TIBMOL, Berlin, Germany).

   Taqman probes are labeled at the 5′-site with a reporter signal using FAM-6-Fluorescence marker and at the 3′-site with a quencher-signal using TAMRA (Applied Biosystems, Darmstadt). The sensor probe of hybridization probes is labeled at the 3′-site with fluorescein. The anchor probe is labeled at the 5′-site with Red640 and at the 3′-site with Phosphate (TIP MOLBIOL).

2. Oligonucleotide primers (Invitrogen).
3. 20-μL capillaries (Roche Diagnostics).
4. Super Script II RT-PCR kit includes:

   $MgS0_4$
   Platin taq®
   2X reaction mix (Invitrogen)

5. Bovine serum albumin (BSA) (Roche Diagnostics).
6. RNAsin (Promega).
7. AmpErase uracil-*N*-glycosylase (UNG) 1U(μL) (Roche Diagnostics).
8. dUTP 25 μ*M* in 250 μL 100 m*M* (Roche, Diagnostics).

2.2.6. FLT3 Length Mutations

1. FAM fluorescent labelled primer (Applied Biosystems).
2. High Fidelity Master Mix (Roche Diagnostics).
3. Rox 500 Internal Standard (Applied Biochemicals).

2.2.7. Cell Lines as Controls for Quantitative PCR

The following cell lines were used:

1. K-562, chronic myeloid leukemia in myeloid blast crisis.
2. Kasumi-1, acute myeloid leukemia [FAB M2 with t(8;21)].
3. MV4-11, acute monocytic leukemia FAB M5 with t (4;11)(q21;q23).
4. NB-4, acute promyelocytic leukemia (FAB M3) with t(15;17).
5. ME-1 with inv(16) AML (all purchased from DSMZ, Braunschweig, Germany).
6. The cells were grown in RPMI 1640 medium (Invitrogen, Heidelberg, Germany) supplemented with 10% fetal bovine serum.
7. All cells were maintained in a humidified 37°C incubator with 5% CO_2 (*see* **Note 2**).

3. Methods

3.1. Preparation of the Samples and RNA Isolation Using the RNeasy Mini Kit

1. All samples should be collected in EDTA tubes. The first step prepares the samples.
2. Use 3–5 mL of blood or bone marrow sample and add erythrocyte-lysis buffer up to a total volume of 50 mL.
3. Place at room temperature for 10 min for erythrocyte lysis.
4. Centrifuge at 1200*g* for 10 min at room temperature.
5. Discard the supernatant and refill up to 25 mL.
6. Centrifuge at 1200*g* for 10 min at room temperature again.
7. According to the size of the obtained pellet, add 350 μL to 2 mL of RLT buffer.
8. Store at –20°C until needed.
9. The RNA isolation using the RNeasy Mini Kit should be performed according Qiagen′s recommendation including the use of Qiashredder (*see* **Note 3**).

3.2. Reverse Transcription and cDNA Synthesis

Prepare the following mixture:

1. Add 8 µL of RNA (2 µg) to 2 µL of Pd (N)6, random hexamers (10 p*M*/µL) on ice.
2. Incubate the mixture at 70°C for 10 min and then on ice for at least 1 min.
3. Prepare reaction master mix:

40 µL total volume	1X (in µL)
10X PCR buffer	5
dNTP	8
M-MLV-RT	0.025
Rnasin	0.25
bdH_2O	27

4. Mix gently and add 10 µL of RNA and Pd(N)6 random hexamers mixture.
5. Incubate the tubes at 37°C for 60 min, heat inactivate at 95°C for 5 min, and then chill on ice.
6. Store at –20°C until use for real-time PCR (*see* **Note 4**).

3.3. Real-time RT PCR protocols

3.3.1. Real-Time RT-PCR Protocol for BCR-ABL Transcripts

1. Detects a2b2 and a2b3 BCR-ABL transcripts (not e1a2).
2. Primers:
   bcr-abl forward: 5′ – ACg TTC CTg ATC TCC TCT gA – 3′ length: 20
   bcr-abl reverse: 5′ – AgA TgC TAC Tgg CCg CTg AA – 3′ length: 20
3. Hybridization probes:

   Anchor probe bcr-abl UP:
   5′ – gCA gAg Tgg Agg gAg AAC ATC Cgg gAg CAg CAg AA—x
   (x = 3′Fluorescein), length: 35

   Sensor probe bcr-abl DOWN:
   5′ - LC Red 640 – AAg TgT TTC AgA AgC TTC TCC CTg AC—p
   (p= 3′Phosphate), length: 26
4. *See* **Table 2** for Lightcycler conditions for bcr-abl.
5. Prepare the following master mix for 1X:

12 µL total volume	1X	Primer FOR	10 p*M*	0.12 *
2X reaction mix	4.8	Primer REV	10 p*M*	0.12 *
50 m*M* $MgSO_4$	0.675	Probe UP	4 p*M*	0.24 *
20 mg/mL BSA	0.3	Probe down	8 p*M*	0.48 *
Super Script Platin Taq	0.3	Rnasin		0.168
dUTP	0.24	UNG		0.6
RNA 1000 ng	4	Mix		8

*Approximate, may vary in volume.

6. Controls (all in duplicate) (*see* **Note 5**): positive control, RNA of the K562 cell line; dilution, 100 ng, 10 ng, 1 ng, 0.1 ng; negative control: bdH_2O.

3.3.2. Real-Time RT-PCR Protocol for AML1/ETO Transcripts

1. Detects the single breakpoint loci of AML1-ETO transcripts
2. Primers :

   AML M2 A1: 5′-AgC CAT gAA gAA CCA gg- 3′ length: 17
   AML M2 E1: 5′- Agg CTg TAg gAg AAT gg- 3′ length: 17

3. Hybridization probes:

   Anchor probe AML-M2 UP:
   5′- ACA ATg CCA gAC TCA CCT gTg gAT gTg AAg ACg C—x
   (x = 3′ Fluorescein), length: 34

   Sensor probe AML-M2 DOWN:
   5′ - LC Red 640 – ATC Tag gCT gAC TCC TCC AAC AAT g—p
   (p= 3′ Phosphate), length: 25

4. *See* **Table 3** for Lightcycler conditions for AML1/ETO:
5. Prepare the following master mix (mix for 1X):

20 μL total volume	1X (in μL)			
2X Reaction Mix	10	Primer FOR	10 p*M*	0.25 *
50 m*M* $MgSO_4$	1.0	Primer REV	10 p*M*	0.25 *
20 mg/mL BSA	0.5	Probe UP	4 p*M*	0.40 *
RNA	6	Probe down	8 p*M*	0.80 *
UNG	14	Super Script Pla Taq		0.50
dUTP		Rnasin		0.28
		Mix		14

   *Approximate, may vary in volume.

6. Controls (all in duplicate) (*see* **Note 6**): positive control, RNA of the Kasumi 1 cell line; dilution, 100 ng, 10 ng, 1 ng, 0.1 ng; negative control, bdH_2O.

3.3.3. Real-Time RT-PCR for CBFβ-MYH11 Transcripts

1. Detects transcripts A, B, C, D, E, and F
2. Primers :

   Forward: 5′-GCAGGCAAGGTATATTTGAAGG- length: 22
   Reverse A: 5′-CTCTTCTCCTCATTCTGCTC- length: 20
   Reverse B: 5′-TTGAGCATTTTGTTGGTCCG – length: 20
   Reverse C: 5′-CTTCCAGAGCTTCCACGGT- length: 19
   Reverse D: 5′- TCCCTGTGACGCTCTCAACT- length: 20
   Reverse E: 5′- GGCCAGGTCTGCGTTCTCT- length: 19
   Reverse F: 5′- TCCTCTTCTCCTCATTCTGCTC- length: 22

3. Hybridization probes:

   Anchor probe CBFB UP:
   GGCTCGCTCCTCATCAAACTCCA—x (x = 3′Fluorescein), length: 23

   Sensor probe CBFB DOWN:
   5′ - LC Red 640 – ACAGCCCATACCATCCAGTCTTTGG
   (p= 3′Phosphate), length: 25

4. *See* **Table 4** for Lightcycler conditions for CBFβ.

Table 2
Lightcycler™ Conditions for BCR-ABL

	Temperature	Incubation time	Temperature transition	Secondary target	Step size	Step delay cycles	Acquisition
Reverse transcription	55°C	20 min	20.00	0	0.0	0	none
Initial denaturation	95°C	30 s	20.00	0	0.0	0	none
denaturation	95°C	1 s	20.00	0	0.0	0	none
Annealing	56°C	15 s	20.00	0	0.0	0	single
Extension	72°C	18 s	2.00	0	0.0	0	none
Cooling	4°C	1 min	20.00	0	0.0	0	none

50 cycles

Table 3
Lightcycler™ Conditions for AML1/ETO

	Temperature	Incubation time	Temperature transition	Secondary target	Step size	Step delay cycles	Acquisition
Reverse transcription	95°C	20 min	20.00	0	0.0	0	none
Initial denaturation	95°C	30 s	20.00	0	0.0	0	none
denaturation	95°C	1 s	20.00	0	0.0	0	none
Annealing	52°C	15 s	20.00	0	0.0	0	single
Extension	72°C	18 s	2.00	0	0.0	0	none
Cooling	4°C	1 min	20.00	0	0.0	0	none

45 cycles

5. Prepare the following master mix for 1X:

20 μL total volume	1X (in μL)	Primer FOR	50 p*M*
2X Reaction Mix	4.8	Primer REV	50 p*M*
50 m*M* $MgSO_4$	0.675	Probe UP	25 p*M*
20 mg/mL BSA	0.3	Probe down	25 p*M*
$MgCl_2$	4 m*M*	LC FastStart DNA Master Hybridization Probes	2 μL
cDNA	2 μL		
Mix			

6. Controls (all in duplicate) (*see* **Note 7**):

   Positive control, RNA of the ME-1 cell line;
   Dilution, 100 ng, 10 ng, 1 ng, 0.1 ng;
   Negative control, bdH_2O.

3.3.4. Real-Time RT-PCR for PML-RARA Transcripts

1. Detects S and L transcripts.
2. Primers:

   AML – M3 – L – forward :
   5′ - gTC TTC CTg CCC AAC AgC AAC C – 3′, length: 22
   AML – M3 – S –forward :
   5′ -AgC TCT TgC ATC ACC CAg ggg A– 3′, length: 22
   AML – M3 – reverse:
   5′ - CTC ACA ggC gCT gAC CCC ATA gT – 3′, length: 23

3. Hybridization probes:

   AML – M3 -UP:
   5′- TgA AgA gAT AgT gCC CAg CCC TCC CTC - x,
   (x = 3′Fluorescein), length: 27

   AML– M3- DOWN:
   5′ - LC Red 640 – CCA CCC CCT CTA CCC CgC ATC TAC A - p ,
   (p= 3′Phosphate), length: 25

4. *See* **Table 5** for Lightcycler conditions for PML-RARA.
5. Prepare the following master mix for 1X:

20 μL total volume	1X (in μL)	Primer L FOR	20 p*M*	0.094 *
2X Reaction Mix	10	Primer S FOR	20 p*M*	0.076 *
50 m*M* $MgSO_4$	0.16	Primer REV	20 p*M*	0.078 *
20 mg/mL BSA	0.5	Probe UP	10 p*M*	0.5 *
Super Script Platin Taq	0.5	Probe down	10 p*M*	0.5 *
Rnasin	0.28	dUTP		0.2
UNG	0.5	bdH_2O		0.6
RNA	6	Master mix		16

*Approximate, may vary in volume.

6. Controls (all in duplicate) (*see* **Note 8**): positive control, RNA of the NB-4 cell line; dilution, 100 ng, 10 ng, 1 ng, 0.1 ng; negative control, bdH_2O.

Table 4
Lightcycler™ Conditions for CBFβ

	Temperature	Incubation time	Temperature transition	Secondary target	Step size	Step delay cycles	Acquisition
Reverse transcription	55°C	20 min	20.00	0	0.0	0	none
Initial denaturation	95°C	30 s	20.00	0	0.0	0	none
Denaturation	95°C	10 s	20.00	0	0.0	0	none
Annealing	64°C	10 s	20.00	0	0.0	0	single
Extension	72°C	26 s	20.00	0	0.0	0	none
Cooling	4°C	1 min	20.00	0	0.0	0	none

45 cycles

Table 5
Lightcycler™ Conditions for PML-RARA

	Temperature	Incubation time	Temperature transition	Secondary target	Step size	Step delay cycles	Acquisition
Reverse transcription	55°C	20 min	20.00	0	0.0	0	none
Initial denaturation	95°C	30 s	20.00	0	0.0	0	none
Denaturation	95°C	10 s	20.00	0	0.0	0	none
Annealing	60°C	10 s	20.00	0	0.0	0	single
Extension	72°C	30 s	2.00	0	0.0	0	none
Cooling	4°C	1 min	20.00	0	0.0	0	none

45 cycles

3.3.5. Real-Time PCR for GAPDH as Housekeeping Gene

1. Primers:
   GAPDH forward: 5′ - TTC ACC ACC ATg gAg AAg gCT – 3′, length: 21
   GAPDH reverse: 5′ - ATg gCA Tgg ACT gTg gTC ATg – 3′, length: 21
2. Hybridization probes:
   Anchor probe GAPDH UP:
   5′- ATC ATC AgC AAT gCC TCC TgC ACC ACC AAC TgC T—x;
   (x = 3′Fluorescein), length: 34
   Sensor probe GAPDH DOWN:
   5′ - LC Red 640 – AgC ACC CCT ggC CAA ggT CAT CCA T – p;
   (p= 3′Phosphate), length: 25
3. *See* **Table 6** for Lightcycler conditions for GAPDH.
4. Prepare the following master mix for 1X:

12 μL total volume	1X in μL	Primer FOR	10 p*M*	0.12 *
2X Reaction Mix	4.8	Primer REV	10 p*M*	0.12 *
50 m*M* $MgSO_4$	0.675	Probe UP	4 p*M*	0.24 *
20 mg/mL BSA	0.3	Sonde down	8 p*M*	0.48 *
Super Script Pla Taq	0.3	Rnasin		0.168
dUTP	0.24	UNG		0.6
RNA	4	Master mix		8

*Approximate, may vary in volume.

5. Controls (all in duplicate) (*see* **Note 9**): positive control, RNA of the K562 cell line;dilution, 100 ng, 10 ng, 1 ng, 0.1 ng; negative control, bdH_2O.

3.3.6. Real-Time RT-PCR Protocol for WT1

1. Detects the transcripts in aberrantly over-expressed in acute and myeloid leukemia.
2. Primers:
   WT 1 forward: 5′ - CgC TAT TCg CAA TCA ggg TTA C – 3′, length: 22
   WT 1 reverse: 5′ - ATg ggA TCC TCA TgC TTg AAT g – 3′, length: 22
3. TaqMan™-Probe:
   WT 1 Probe: 6 FAM – Cgg TCA CCT TCg ACg ggA CgC xT – pH, length: 22
4. See **Table 7** for Lightcycler conditions for WT1.
5. Prepare the following master mix for 1X:

12 μL total volume	1X (in μL)	Primer FOR	20 p*M*	0.08 *
2X Reaction Mix	4.8	Primer REV	20 p*M*	0.08 *
50 m*M* $MgSO_4$	0.675	Taqman probe		0.24 *
UNG	0.6	Super Script	0.3	
20mg/mL BSA	0.3	dUTP	0.24	
Rnasin	0.168	RNA	4	
Mix	7.5			

*Approximate, may vary in volume.

6. Controls (all in duplicate) (*see* **Note 10**): positive control, RNA of the K562 cell line; dilution, 100 ng, 10 ng, 1 ng, 0.1 ng; negative control, bdH_2O.

Table 6
Lightcycler™ Conditions for GAPDH

	Temperature	Incubation time	Temperature transition	Secondary target	Step size	Step delay cycles	Acquisition
Reverse Transcription	55°C	20 min	20.00	0	0.0	0	none
initial denaturation	95°C	30 s	20.00	0	0.0	0	none
denaturation	95°C	1 s	20.00	0	0.0	0	none
annealing	56°C	15 s	20.00	0	0.0	0	single
extension	72°C	18 s	2.00	0	0.0	0	none
cooling	4°C	1 min	20.00	0	0.0	0	none

50 cycles

Table 7
Lightcycler™ Conditions for WT1

	Temperature	Incubation time	Temperature transition	Secondary target	Step size	Step delay cycles	Acquisition
Reverse transcription	55°C	20 min	20.00	0	0.0	0	none
initial denaturation	95°C	30 s	20.00	0	0.0	0	none
denaturation	95°C	1 s	20.00	0	0.0	0	none
Annealing	56°C	15 s	20.00	0	0.0	0	single
Extension	72°C	18 s	2.00	0	0.0	0	none
Cooling	4°C	1 min	20.00	0	0.0	0	none

45 cycles

3.3.7. FLT3 Length Mutation Detection by Genetic Analyzer

1. Detectable in up to 30% of all acute myeloid leukaemia.
2. Primers:

   FLT3 11F: 5′ - 6-FAM- gCAATTTAggTATgAAAgCCAgC– 3′, length: 23
   FLT3 12R: 5′ - CTTCAgCATTTTgACggCAACC – 3′, length: 22

3. PCR-Program: FLT3 – A :

Initial denaturation:	94°C	2 min	
Denaturation :	94°C	30 s	
Annealing	60°C	1 min	30 cycles
Extension :	72°C	1.15 min	
Final extension:	68°C	45 min	
Cooling:	4°C		

4. Prepare the following master mix for 1X:

50 µL total volume	1X (in µL)			
1X High Fidelity Master Mix	25	FLT3 – 11 Fam	15 p*M*	0.75 *
bdH_2O	19.18	FLT3 – 12 R	15 p*M*	0.07 *
Master mix	45	100 ng DNA		

*Approximate, may vary in volume.

5. ABI Prism Genetic Analyzer 310 conditions:

Polymer	POP 4	Injection time in	1 s
Capillary	47 cm	Injection	10.0 kV
Dye Set	DS 30 (Matrix)	Run	15.0 kV
Filter Set	D (GS STR POP 4- 1mL)	Run	55°C
Size STD	SSTD 500 Rox FLT3	Run (time)	66 min

6. Preparation of samples: mix 1 µL of PCR product with 8.5 µL bdH_2O and 0.5 µL ROX500 internal size standard.
7. Prepare denaturation of the samples at 95°C for 3 min and then 4°C for 3 min.
8. Results: data can be analyzed with the ABI Prism Genetic Analyzer 310 including Genemapper v.3.10 software. The wild type has 329 bp, whereas the peaks (one or more) of length mutations can be detected >329 bp. For controls, the cell line MV 4-11, which shows peaks at 359 bp, can be used (*see* **Note 11**).

3.3.8. Multiplex Nested Qualitative RT- PCR for 11Q23 Chromosomal Aberrations [t(6;11), t(9;11) and t(11;19)]

1. Detects MLL/AF4, MLL/AF6, MLL/AF9, and MLL/ENL transcripts
2. Primers:

   P1: 5′ - CCT gAA TCC AAA CAg gCC ACC ACT – 3′, length: 24
   P2: 5′ - CTT CCA ggA AgT CAA gCA AgC Agg – 3′, length: 24
   P3: 5′ - gTC ACT gAg CTg AAg gTC gTC TTC g – 3′, length: 25
   P4: 5′ - AgC ATg gAT gAC gTT CCT TgC TgA g – 3′, length: 25

Primers continued on next page

2. Primers: *(Continued from previous page)*

P5:	5′ - TCC AAT TCA gTT gTA CAA CTA gAg g – 3′,	length: 25
P6:	5′ - CCA ATC TTC TTT CTC CgC TgA CAT g – 3′,	length: 25
P7:	5′ - TCT gAT TTg CTT TgC TTT ATT ggA C – 3′,	length: 25
P8:	5′ - CgT gAT gTA ggg gTg AAg AAg CAg – 3′,	length: 24
P9:	5′ - CCA CgA AgT gCT ggA TgT CAC AT – 3′,	length: 23
P10:	5′ - Cgg ACA AAC ACC ATC CAg TCg Tg – 3′,	length: 23

3. First step: cDNA synthesis (*see* **Subheading 3.2.**).
4. Nested PCR program as follows:

Program: 11Q23 1. Step:			Program: 11Q23 2. Step:		
94°C	1:00 min		94°C	1:00 min	
60°C	2:00 min	35X	60°C	2:00 min	25X
74°C	2:00 min		74°C	2:00 min	
4°C			4°C		

5. Prepare the following master mix for 1X (1. step):

25µL total volume		1X (in µL)
bdH_2O		30.92
dNTP		8.0
10X PE buffer		5.0
Primer 1	10 p*M*	
Primer 3	10 p*M*	
Primer 5	10 p*M*	
Primer 7	10 p*M*	
Primer 9	10 p*M*	
PE Taq		0.4
cDNA		5 µL
Mix		45 µL

6. Master Mix for 2. PCR (2. step):

25 µL volume		1X (in µL)
bdH_2O		30.9
dNTP		8.0
10X PE buffer		5.0
Primer 2	10 p*M*	
Primer 4	10 p*M*	
Primer 6	10 p*M*	
Primer 8	10 p*M*	
Primer 10	10 p*M*	
PE Taq		0.4
cDNA		5 µL
Mix		45 µL

7. Controls: cDNA of cell line MV4-11.
8. Visualize PCR-products on a 1. 5% agarose gel stained with ethidium bromide (*see* **Note 12**).

4. Notes

1. All quantitative PCR protocols here are designed for use of a Lightcycler device.
2. For positive controls, RNA of a specific cell line was used. The use of plasmid as controls is also possible, but are associated with a slightly more crossover decontaminations.
3. In order to avoid carry-over decontamination, the recommendations of Kwok and Higuchi should be considered strictly ***(28)***. RNA or DNA isolation should be performed at different places. Use positive-displacement pipets to avoid inaccuracies in pipetting.
4. Reverse transcription of mRNA to cDNA should be done with random hexamers (not with oligo-dT). Most of the above shown real-time RT-PCR protocols do not have a separate cDNA-synthesis step here.
5. This is a one-step real-time RT-PCR using mRNA as a template ***(5)***. The master mix contains only 4.8 µL instead of 6 µL of a 2X reaction mix in 12 µL total PCR volume, which is absolutely sufficient.

   It is of utmost importance to realize that results of real-time PCR for one sample can vary between laboratories or even between individuals, who perform the test as shown in internal laboratory quality assurance tests ***(11)***. To obtain accurate and reliable results, all standard dilutions and patient samples should be tested in duplicate. For quantification, the average value of both duplicates should be used. Samples that had a more than 100-fold difference for both housekeeping gene values should be excluded from further analysis. Samples that test negative for the housekeeping gene but positive for the target gene should be excluded from quantification. A negative assay requires the absence of the target PCR product as well as no amplification of the "blank" sample and a positive amplification of the positive control, but a successful housekeeping gene PCR amplification. The expression of the target gene is given as the quotient of the target gene and housekeeping gene. The housekeeping gene should also be related to a standard curve derived from a serial dilution of plasmids or RNA of a target specific leukemic cell line into water.
6. UNG is added in the reaction to prevent the re-amplification of carry-over PCR products by removing any uracil incorporated into amplicons (*see also* **refs. *12–14***).
7. Please note that template is cDNA for this real-time PCR here. There are six different CBFβ-MYH11 transcripts commonly found in AML with inv(16). Type A is most commonly found up to 80%. Further, the control cell line ME-1 expresses only type A; therefore, it is often difficult to perform quantitative PCR for the other subtypes without specific control dilutions. For each subtype, a separate PCR is needed with the specific reverse primers and the common forward primer. The use of SYBR® green instead of hybrizidation probes is not recommended because of the high risk of contamination, which can shut down every laboratory for a longer period of time (*see also* **refs. *15–19***).
8. This real-time RT-PCR is again a single-step PCR using RNA as template. It detects the fusion transcripts S and L of PML-RARA in a single run, but not the variant type (*see also* **refs. *12*** and ***20***).

9. As housekeeping genes, GAPDH, G6PDH, Abelson (ABL), β-2-microglobulin (B2M), and β-glucorinase (GUS) are most commonly used. To avoid pseudo-genes and maintain a stability in the level of expression of mRNA in each cell B2M, ABL and β-glucuronidase were recently recommended by the Europe Against Cancer (EAC) program *(9)*. Nevertheless, most laboratories continue to use GAPDH as their housekeeping genes *(5)* (for quantification, *see also* **Note 1**).
10. The use of WT1 as a MRD marker is still discussed controversial and should be used only if no other specific MRD marker is available. This real-time RT-PCR uses a Taqman probe. K562 can be used as a control cell line (*see also* **refs. *21–23***).
11. Use only the high-fidelity master mix kit (from Roche Diagnostics) for PCR when a genetic analyzer is used. Results showing more than a single PCR-product of 329 bp indicates that a FLT3 mutation is detected (*see also* **refs. *24–26***).
12. This is a very robust nested multiplex PCR, which uses cDNA as template and detects the most common mixed-lineage leukemia rearrangement genes, MLL/AF4, MLL/AF6, MLL/AF9, and MLL/ENL (*see also* **ref. *27***).

References

1. Elmaagacli, A. H., Becks H. W., Beelen, D. W., et al. (1995) Detection Of minimal residual disease And persistence of host-type hematopoiesis: a study in 28 patients after sex-mismatched, non T-cell-depleted allogeneic bone marrow transplantation for philadelphia- chromosome positive chronic myelogenous leukemia. *Bone Marrow Transplant.* **16,** 823–829.
2. Mackinnon, S., Barnett, L., Heller, G., and O'Reilly, R. J. (1994) Minimal residual disease is more common in patients who have mixed T-cell chimerism after bone marrow transplantation for chronic myelogenous leukemia. *Blood* **83,** 3409–3416.
3. Kolb, H., Mittermüller, J., Clemm, C., et al. (1990) Donor leucocyte transfusions for treatment of recurrent chronic myelogenous leukemia in marrow transplant patients. *Blood* **76,** 2462–2465.
4. Elmaagacli, A. H., Beelen, D. W., Opalka, B., Seeber, S., and Schaefer, U. W. (1999) The risk of residual molecular and cytogenetic disease in patients with Philadelphia-chromosome positive first chronic phase chronic myelogenous leukemia is reduced after transplantation of allogeneic peripheral blood stem cells compared to bone marrow. *Blood* **94,** 384–389.
5. Elmaagacli, A. H., Freis, A., Hahn, M., et al. (2001) Estimate the relapse stage in chronic myeloid leukaemia patients after allogeneic stem cell transplantation by the amount of BCR-ABL fusion transcripts detected using a new real-time polymerase chain reaction method. *Br. J. Haematol.* **113,** 1072–1075.
6. Reading, C. L., Estey, E. H., Huh, Y. O., et al. (1996) Expression of unusual immunophenotype combinations in acute myelogenous leukemia. *Blood* **81,** 3083–3090.
7. Macedo, A., San Miguel, J. F., Vidriales M. B., et al. (1996) Phenotypic changes in acute myeloid leukaemia: implications on the detection of minimal residual disease. *J. Clin. Pathol.* **49,** 15–18.

8. Bernell, P., Arvidsson, I., Jacobsson, B., and Hast, R. (1996) Fluorescence in situ hybridzation in combination with morphology detects minimal residual disease in remission and heralds relapse in acute leukaemia. *Br. J. Haematol.* **95,** 666–672.
9. Weisser, M., Haferlach, T., Schoch, C., Hiddemann, W., and Schnittger, S. (2003) The use of housekeeping genes for real-time PCR based quantification of fusion gene transcripts in acute myeloid leukemia. *Leukemia* **17,** 2474–2486.
10. Holland, P. M., Abramson, R. D., Watson, R., and Gelfand, D. H. (1991) Detection of specific poylmerase chain reaction product utilizing the 5′-3′ exonuclease activity of Thermus aquaticus DNA polymerase. *Proc. Natl. Acad. Sci. USA* **88,** 7276–7280.
11. Burmeister, T., Maurer, J., Aivado, M., et al. (2000) Quality assurance in RT-PCR-based BCR/ABL diagnostics—results of an interlaboratory test and a standardization approach. *Leukemia* **14,** 1850–1856.
12. Schnittger, S., Weiser, M., Schoch, C., Hiddemann, W., Haferlach, T., and Kern, W. (2003) New score predicting for prognosis in PML-RARA+, AML1-ETO+, or CBFB-MYH11+ acute myeloid leukemia based on quantification of fusion transcripts. *Blood* **102,** 2746–2755.
13. Krauter, J., Wattjes, M. P., Nagel, S., et al. (1999) Real-time RT-PCR for the detection and quantification of AML1/MTG fusion transcripts in t(8;21)-positive AML patients. *Br. J. Haematol.* **107,** 80–85.
14. Elmaagacli, A. H., Beelen, D. W., Kroll, M., et al. (1997) Detection of AML1/ETO fusion transcripts in patients with t(8;21) acute myeloid leukemia after allogeneic bone marrow transplantation or peripheral blood progenitor cell transplantation. *Blood* **90,** 3230–3231.
15. Guerrasio, A., Pilatrino, C., De Micheli, D., et al. (2002) Assessment of minimal residual disease (MRD) in CHFbeta/MYH11-positive acute myeloid leukemias by qualitative and quantitative RT-PCR amplifikation of fusion transcripts. *Leukemia* **16,** 1176–1181.
16. Krauter, J., Hoellge, W., Wattjes, M. P., et al. (2001) Detection and quantification of CBFB/MYH11 fusion transcripts in patients with inv(16)-positive acute myeloblastic leukaemia by real-time RT-PCR. *Genes Chromosom. Cancer* **30,** 342–348.
17. Marcucci, G., Caliguri, M. A., Dohner, H., et al. (2001) Quantification of CBFbeta/MYH11 fusion transcript by real time RT-PCR in patients with INV(16) acute myeloid leucemia. *Leukemia* **15,** 1072–1080.
18. Buonamici, S., Ottaviani, E., Testoni, N., et al. (2002) Real-time quantification of minimal residual disease in inv(16)-positive acute myeloid leukaemia may indicate risk for clinical relapse and may identify patients in a curable state. *Blood* **99,** 443–449.
19. Elmaagacli, A. H., Beelen, D. W., Kroll, M., Trzensky, S., Stein, C., and Schaefer, U. W. (1998) Detection of CBFB/MYH11 fusion transcripts in patients with inv(16) acute myeloid leukemia after allogeneic bone marrow or peripheral blood progenitor cell transplantation. *Bone Marrow Transplant.* **21,** 159–166.
20. Gallagher, R. E., Yeap, B. Y., et al. (2003) Quantitative real-time RT-PCR analysis of PML-RARA alpha mRNA in acute promyelocytic leukaemia: Assess-ment

of prognostic significance in adult patients from intergroup protocol 0129. *Blood* **101,** 2521–2528.
21. Ogawa, H., Tamaki, H., Ikegame, K., et al. (2003) The usefilness of monitoring WT1 gene transcripts for the prediction and management of relapse following allogeneic stem cell transplantation in acute type leukaemia. *Blood* **101**, 1698–1704.
22. Cilloni, D., Gottardi, E., Fava, M., et al. (2003) Uselfulness of quantitative assssment of the WT1 gene transcript as a marker for minimal residual disease detection. *Blood* **102,** 773–774.
23. Elmaagacli, A. H., Beelen, D. W., Trenschel, R., and Schaefer, U. W. (2000) The detection of wt-1 transcripts is not associated with an increased leukemic relapse rate in patients with acute leukemia after allogeneic bone marrow or peripheral blood stem cell transplantation. *Bone Marrow Transplant.* **25,** 91–96.
24. Vaughn, C. P. and Elenitoba-Johnson, K. S. J. (2004) High-resolution melting analysis for detection of internal tandem duplications. *J. Mol. Diagn.* **6,** 211–216.
25. Schnittger, S., Schoch, C., Dugas, M., et al. (2002) Analysis of FLT3 length mutations in 1003 patients with acute myeloid leukemia: correlation to cytogenetics, FAB subtype, and prognosis in the AMLCG study and usefulness as a marker for the detection of minimal residual disease. *Blood* **100,** 59–66.
26. Yokota, S., Kiyoi, H., Nakao, M., et al. (1997) Internal tandem duplications of the FLT3 gene is preferentially seen in acute myeloid leukemia and myelodysplastic syndrome among various hematological malignancies. A study on a large series of patients and cell lines. *Leukemia* **11,** 1605–1609.
27. Repp, R., Borkhardt, A., Haupt, E., et al. (1995) Detection of four different 11q23 chromosomal abnormalities by multiplex-PCR and fluorescent-based automatic DNA-fragment analysis. *Leukemia* **9,** 210–216.
28. Kwok, S. and Higuchi, R. (1989) Avoiding false positives with PCR. *Nature* **339,** 237–238.

13

Molecular Methods Used for The Detection of Autologous Graft Contamination in Lymphoid Disorders

Paolo Corradini, Matteo G. Carrabba, and Lucia Farina

Summary

Intensified treatments aimed at maximal tumor reduction are an important therapeutic option for patients affected by B-cell malignancies. The possibility of obtaining a relevant number of clinical complete remissions after these treatments prompted the application of molecular techniques for the detection of extremely low numbers of residual malignant cells. These cells can be present either in the stem cell graft or, during the follow-up, in the bone marrow of patients attaining a clinical complete remission. The most sensitive and widely used techniques for minimal residual disease (MRD) assessment are those based on the PCR method. These methods allow the detection of autologous graft contamination and the identification of patients at high risk of disease recurrence by means of post-transplant MRD monitoring. In this setting, quantitative PCR assays can evaluate the kinetics of tumor clone growth in complete remission (CR) patients showing a persistence of PCR detectable tumor cells with standard qualitative methods.

Key Words: Polymerase chain reaction; minimal residual disease; lymphomas; leukemia.

1. Introduction

In the last two decades, the development of high-dose chemotherapy (HD-CT) and autologous stem cell transplantation (auto-SCT) protocols have changed the therapeutic approach for some B-cell non-Hodgkin lymphomas (NHLs). Particularly patients with low-grade NHL, incurable pathologies by conventional treatments benefit remarkably from these programs in terms of disease response, disease-free survival (DFS), and overall survival (OS) *(1,2)*. Because most of these patients have bone marrow infiltration, a major concern with autologous stem cells is the risk of malignant cell reinfusion *(3)*. The PCR

From: *Methods in Molecular Medicine, vol. 134: Bone Marrow and Stem Cell Transplantation*
Edited by: M. Beksac © Humana Press Inc., Totowa, NJ

technique uses DNA sequences that are constantly detectable in the neoplastic population and not present in the normal hematopoietic compartment. In B-cell malignancies, tumor-specific chromosomal translocations and clonal rearrangements of immunoglobulin heavy chain (IgH) genes can be used as molecular markers. PCR detection of tumor cells is clinically useful for (1) assessment of residual malignant cells in the graft, and (2) the molecular monitoring of patients in clinical complete remission (CR) after transplantation. It has been shown that patients whose marrow graft contained PCR-detectable lymphoma cells after immunological ex vivo purging had an increased incidence of relapse after auto-SCT ***(4,5)***. Besides, molecular monitoring of serial bone marrow samples obtained after transplant was performed to assess whether the auto-SCT approach was able to induce a molecular remission ***(6)***. The persistence or reappearance of PCR-detectable lymphoma cells has been correlated with a worse DFS. In the setting of indolent lymphoma patients receiving HD-CT and auto-SCT it has been demonstrated that the collection of exclusively PCR-positive harvests and the persistence of PCR positivity following autografting can be considered a strong prognostic factor for disease recurrence ***(1,7)***. Among low-grade lymphomas, the different disease entities have been associated with different responses to these high dose therapies. For example, a sizeable proportion of follicular lymphoma (FCL) patients can have PCR-negative harvests and post-transplant molecular remission (MR); on the contrary, the MRs are very rare in mantle cell lymphomas (MCLs). It is important to point out that the recent addition of the humanized anti-CD20 (Rituximab) to high-dose programs has increased the number of patients attaining the molecular remission ***(8)***.

The role of auto-SCT for chronic lymphocytic leukemia (CLL) has not been clearly defined yet. Some studies have showed an improvement in progression-free and overall survival ***(9–11)***; nevertheless the curves did not reach a plateau phase because of a continuous relapse rate ***(12)***. Also in this disease, MRD positivity after auto-SCT has shown to be a predictor of relapse: PCR-positive patients, and PCR-negative patients that become positive, inevitably relapse ***(13)***. In vitro purging of stem cell harvests has failed to reach the PCR negativity in most of the patients and to demonstrate any improvement of the DFS ***(14)***. Studies with monoclonal antibodies (Campath 1H or Rituximab) as an in vivo purging strategy are ongoing.

Acute lymphoblastic leukemia (ALL) is another hematological malignancies for which MRD monitoring can help to identify patients at high risk of relapse after induction chemotherapy. Because of the possibility of clonal evolution of leukemia cells, usually two or more independent molecular markers are needed to prevent the detection of false-negative results ***(15,16)***. Several prospective clinical studies in childhood ALL have shown that MRD levels can identify groups at high, intermediate, and low risk of relapse, especially when early remission time points are analyzed ***(16–18)***. However, because auto-SCT is very rarely used in ALL, this chapter will not address the issue of MRD monitoring in this disease.

In the setting of molecular monitoring, a number of quantitative and semi-quantitative PCR methods to evaluate the tumor burden have been investigated. Among them the real-time quantitative PCR is a recently introduced technique that allows the determination of residual tumor, overcoming most of the limitations associated with previous quantitative or semi-quantitative PCR strategies ***(19)***. The main advantages associated with the real-time PCR method are (1) the measurement of target DNA copy number is made in the PCR tube as the reaction proceeds without the need for post-PCR sample processing; (2) it greatly reduces the occurrence of false-positive results by adding the specificity of a probe to the one given by primers; and (3) it has a relatively wide dynamic range, allowing the accurate quantification of samples over a wide range of tumor contamination with a sensitivity of 10^{-6}–10^{-4}. Besides the wide use of quantitative PCR in ALL patients, this technique has been employed also for the detection of autologous graft contamination in lymphoma patients. In this setting, a relationship between tumor burden of stem cell harvests and successful purging of PCR-detectable disease following ex vivo manipulation was shown ***(20)***.

Currently, a molecular marker can be identified in approx 75% of B-cell lymphoma patients. The three most widely used clonal markers are derived from t(14;18), t(11;14), or the IgH gene rearrangements.

2. Materials

2.1. Mononuclear Cells Isolation

1. Phosphate-buffered saline (PBS) 1X solution without calcium/magnesium or sodium chloride solution 0.9%.
2. Ficoll-Hypaque gradient (Sigma-Aldrich, Steinheim, Germany).

2.2. DNA Extraction

1. DNAzol reagent (Invitrogen, Carlsbad, CA).
2. Ethanol 100% stored at –20°C.
3. Ethanol 70% stored at –20°C.
4. Sterile water or 8 m*M* NaOH or 0.1X TE Buffer.

2.3. Qualitative PCR

1. dNTPs 2 m*M* (0.5 m*M* of each deoxyadenosine triphosphate [dATP], deoxycytidine triphosphate [dCTP], deoxyguanosine triphosphate [dGTP], and deoxythymidine triphosphate [dTTP]).

Table 1
Primers for Minimal Residual Disease Analysis of Bcl-2 MBR/mcr and Bcl-1 by Qualitative PCR

Forward primers		
• Bcl-2 MBR first PCR	MBR-2	5′-CAGCCTTGAAACATTGATGG-3′
• Bcl-2 MBR second PCR	MBR-3	5′-ATGGTGGTTTGACCTTTAG-3′
• Bcl-2 mcr first PCR	mcr-2	5′-CGTGCTGGTACCACTCCTG-3′
• Bcl-2 mcr second PCR	mcr-3.1	5′-CCTGGCTTCCTTCCCTCTG-3′
• Bcl-1 first PCR	P2	5′-GAAGGACTTGTGGGTTGC-3′
• Bcl-1 second PCR	P4	5′-GCTGCTGTACACATCGGT-3′
Reverse primers		
• First PCR	JH3	5′-ACCTGAGGAGACGGTGACC-3′
• Second PCR	JH4	5′-ACCAGGGTCCCTTGGCCCCA-3′

MBR, major breakpoint cluster region; mcr, minor cluster region.

2. 10X Buffer Taq polymerase (500 m*M* KCl, 100 m*M* Tris-HCl, pH 8.0, 1% [w/v] gelatin) (Promega, Madison, WI).
3. 10X Buffer AmpliTaqGold (Applied Biosystems, Warrington, UK).
4. 15 m*M* and 20 m*M* $MgCl_2$ Solution.
5. Taq polymerase 5 U/µL (Promega, Madison, WI).
6. AmpliTaq Gold 5 U/µL (Applied Biosystems, Warrington, UK).
7. Forward and reverse primers (20 pmol/µL) (*see* **Tables 1** and **2**).
8. Sterile water or DNase/RNase free water.
9. 2% agarose gel.
10. 1% low melting point (LMP) agarose gel.
11. 12% polyacrylamide gel (19:1 acrylamide: bisacrylamide).
12. Ethidium bromide (10 mg/mL) (**caution:** carcinogenic).
13. Loading buffer (15% glycerol and 0.25% xylene cyanol or 0.25% bromophenol blue).
14. Buffer 50X TAE: 242 g Tris base, 57.1 mL glacial acetic acid, 100 mL 0.5 *M* EDTA pH 8.0, H_2O to 1 L.
15. Molecular weight marker (40 µg ΦX-174-Hae III digested DNA, 80 µg – λ Hind III digested DNA); adjust volume to 1 mL with 0.1X TE (10 m*M* Tris-HCl, 1 m*M* EDTA).
16. QIAquick gel extraction kits (Qiagen, Milan, Italy).
17. pGEM®-T Easy Vector System for TA cloning (Promega, Madison, WI).; *Escherichia coli* competent cells (JM 109) (Promega, Madison, WI); Agar plates; Miniprep kit (Promega, Madison, WI).

Reagents stock for the PCR reaction and molecular weight marker are stored at –80°C. Reagents aliquots are stored at –20°C; molecular weight marker and loading buffer aliquots are stored at 4°C.

Table 2
Primers for Immunoglobulin Heavy Chain Gene VDJ Rearrangement Amplification

Forward primers			
FR1 region-derived primers			
VH/FS		VH/D	
• VH1/FS	5′-CAGGTGCAGCTGGTGCA(C/G)(A/T)CTG-3′	VH1/D	5′-CCTCAGTGAAGGTCTCCTGCAAGG-3′
• VH2/FS	5′-CAG(A/G)TCACCTTGAAGGAGTCTG-3′	VH2/D	5′-TCCTGCGCTGGTGAAAGCCACACA-3′
• VH3/FS	5′-CAGGTGCAGCTGGTG(G/C)AGTC(C/T)G-3′	VH3/D	5′-GGTCCCTGAGACTCTTCCTGTGCA-3′
• VH4a/FS	5′-CAG(C/G)TGCAGCTGCAGGAGTC(C/G)G-3′	VH4a/D	5′-TCGGAGACCCTGTCCCTCACCTGCA-3′
• VH4b/FS	5′-CAGGTGCAGCTACA(A/G)CAGTGGG-3′	VH4b/D	5′-CGCTGTCTCTGGTTACTCCATCAG-3′
• VH5/FS	5′-GAGGTGCAGCTG(G/T)TGCAGTCTG-3′	VH5/D	5′-GAAAAAGCCCGGGGAGTCTCTGAA-3′
• VH6/FS	5′-CAGGTACAGCTGCAGCAGTCAG-3′	VH6/D	5′-CCTGTGCCATCTCCGGGGACAGTG-3′
Leader region-derived primers			
VH/LS		VH/L	
• VH1/LS	5′-CTCACCATGGACTGGACCTGGAG-3′	VH1/L	5′-CCATGGACTGGACCTGGAGG-3′
• VH2/LS	5′-ATGGACATACTTTGTTCCACGCTC-3′	VH2/L	5′-ATGGACATACTTTGTTCCAG-3′
• VH3/LS	5′-CCATGGAGTTTGGGCTGAGCTGG-3′	VH3/L	5′-CCATGGAGTTTGGGCTGAGC-3′
• VH4/LS	5′-ACATGAAACA(C/T)CGTGGTTCTTCC-3′	VH4/L	5′-ATGAAACACCTGTGGTTCTT-3′
• VH5/LS	5′-ATGGGGTCAACCGCCATCCTCCG-3′	VH5/L	5′-ATGGGGTCAACCGCCATCCT-3′
• VH6/LS	5′-ATGTCTGTCTCCTTCCTCATCTTC-3′	VH6/L	5′-ATGTCTGTCTCCTTCCTCAT-3′
• VH7/LS	5′-TTCTTGGTGGCAGCAGCCACA-3′		
FR3 region-derived primers			
FR3.3			
• 5′-ACACGGC(C/T)(G/C)TGTATTACTGT-3′			
Reverse primers			
FR1 and leader VH families → JHD		FR3.3 → JH3.3	
• 5′-ACCTGAGGAGACGGTGACCAGGGT-3′		5′-GTGACCAGGGT(A/G/C/T)CCTTGGCCCCAG-3′	

Table 3
Primers and Probes for Quantitative Analysis of Bcl2-MBR by Real-Time PCR

Bcl-2 MBR	
• Forward primer	5′-CTATGGTGGTTTGACCTTTAGAGAG-3′
• Reverse primer	5′-ACCTGAGGAGACGGTGACC-3′
• Probe	5′ FAM-CTGTTTCAACACAGACCCACCCAGAC-TAMRA 3′
GAPDH	
• Forward primer	5′-CAAAGCTGGTGTGGGAGG-3′,
• Reverse Primer	5′-CTCCTGGAAGATGGTGATGG-3′
• Probe	5′ VIC-CAAGCTTCCCGTTCTCAGCC-TAMRA 3′

MBR, major breakpoint cluster region; GAPDH, glyceraldehyde-3-phosphate dehydrogenase.

2.4. Quantitative PCR

1. TaqMan® Universal PCR Master Mix (Applied Biosystems, Warrington, UK).
2. Forward and reverse primers (20 pmol/µL or according to optimization procedure described in Applied Biosystems Instrument Manual) (*see* **Table 3**).
3. TaqMan probe (10 pmol/µL or according to optimization procedure described in Applied Biosystems Instrument Manual).
4. Sterile water or DNase/RNase-free water. Reagents stock are stored at –80°C. Reagents aliquots are stored at –20°C. TaqMan probes are stored in the dark.

3. Methods

3.1. Mononuclear Cells Isolation

1. Dilute peripheral blood or bone marrow samples with 1X PBS (dilution factor 1:1 for peripheral blood and from 1:2 to 1:5 for bone marrow) (*see* **Notes 1** and **2**).
2. Add 4 mL of Ficoll/Hypaque (Sigma, Steinheim Germany) gradient to a 15-mL tube for a sample volume ≤10 mL or add 10 mL of Ficoll/Hypaque gradient to a 50-mL tube for a sample volume ≥10 mL.
3. Lay the sample on the top of the Ficoll slowly and keeping the tube at 45°.
4. Centrifuge at 400*g* for 25 min without brake.
5. Carefully remove the mononuclear cells from the Ficoll/plasma interface and collect them into a new tube.
6. Complete to 50 mL by adding 0.9% sodium chloride solution and centrifuge 400*g* for 10 min with brake; repeat this step twice.
7. Remove supernatant and resuspend the pellet.
8. Collect the cells in a 1.5 mL sterile plastic microcentrifuge tube.
9. Spin the tube in a microcentrifuge at 10,000*g* for 1 min.
10. Remove supernatant and store at –80°C.

3.2. DNA Extraction

1. Add 1 mL of DNAzol to 10×10^6 frozen cells and gently re-suspend the pellet (*see* **Notes 3** and **4**).
2. Centrifuge at 10,000*g* for 10 min at room temperature.
3. Collect the supernatant in a new tube; do not remove the sediment.
4. Add 0.5 mL (for 1 mL of DNAzol) of 100% ethanol, mix gently by inversion.
5. Remove the DNA precipitate by spooling with a pipette tip and collect it into another tube.
6. Wash with 500 μL of 70% ethanol twice.
7. Remove the DNA precipitate from ethanol and collect into a new 1.5-mL sterile plastic microcentrifuge tube (*see* **Note 5**).
8. Air-dry the DNA by storing open the tube for 5–15 min at room temperature.
9. Dissolve the DNA in sterile water or 8 m*M* NaOH (60–130 μL based on the precipitate dimensions).
10. Store the sample at 4°C.
11. Check the DNA quality and the quantity by means of an ultraviolet spectrophotometer (dilute 2 μL of DNA [e.g., 1:70] with sterile water; check the sample purity by means of 260 nm/280 nm ratio [range 1.7–2] and the DNA concentration as follows: 260 nm value × dilution factor × 50= DNA concentration μg/mL).

3.3. Qualitative PCR

3.3.1. Bcl-2/IgH Rearrangement

FCL is characterized by the translocation t(14;18)(q32;q21), in which the bcl-2 proto-oncogene, located on chromosome 18, is juxtaposed to the IgH locus on chromosome 14 ***(21)***. This rearrangement is detectable in up to 85% of FCL patients and in 25% of diffuse large-cell lymphomas. The translocation occurs most frequently in two molecular sites: the major breakpoint cluster region (MBR), a restriction fragment of 2.8 kb within the 3′ untranslated region of the bcl-2 gene (70% of patients) ***(22)***, and the minor cluster region (mcr), located 20 kb downstream (10–15% of patients) ***(23)***. The clustering of the breakpoints at these two main regions and the availability of consensus IgH joining regions make this an ideal PCR target for the detection of lymphoma cells. Because there is a sequence diversity at the site of the breakpoint, the PCR product length, assessed by gel electrophoresis, dictates the specificity for a given patient.

The PCR reactions described below are used to identify the molecular marker as well as for MRD analysis. They must include:

- A bcl-2/IgH MBR/mcr positive sample to assess the sensitivity;
- Healthy donor polyclonal DNA as a negative sample to assess the specificity;
- A no-template control sample to exclude any contamination.

3.3.1.1. Bcl-2 MBR Nested PCR

First PCR amplification:

1. Reaction mix (× *n* samples + 1) (*see* **Note 6**): dNTPs 2 m*M* 5 µL; 10X Buffer Taq polymerase (Promega, Madison, WI) 5 µL; 20 m*M* $MgCl_2$ 5 µL; MBR-2 (20 pmol/µL) 1 µL; JH3 (20 pmol/µL) 1 µL; Taq polymerase (Promega, Madison, WI) 0.2 µL (*see* **Note 7**); sterile water for a final volume of 50 µL.
2. Aliquot PCR reaction mix in each tube (*see* **Note 8**).
3. Add 0.5–1 µg of DNA sample in each tube except the no-template control sample.
4. Amplification profile: first step × 30 cycles: 94°C 1 min; 55°C 1 min; 72°C 1 min; second step: 72°C 7 min (*see* **Note 9**).

Second PCR amplification:

1. Reaction mix (× *n* samples + 1) (*see* **Note 6**): dNTPs 2 m*M* 5 µL; 10X Buffer (Promega, Madison, WI) 5 µL; 20 m*M* $MgCl_2$ 5 µL; MBR-3 (20 pmol/µL) 1 µL; JH4 (20 pmol/µL) 1 µL; Taq DNA Polymerase (Promega, Madison, WI) 0.2 µL; sterile water for a final volume of 50 µL.
2. Aliquot PCR reaction mix in each tube.
3. Add 5 µL of first PCR product (*see* **Note 10**).
4. Amplification profile: first step × 30 cycles: 94°C 1 min; 58°C 1 min; 72°C 1 min; second step: 72°C 7 min (*see* **Note 9**).
5. Run 10 µL of PCR product (with 1 µL of loading buffer) and 10 µL of molecular weight marker (with 1 µL of loading buffer) on a 2% agarose gel stained with ethidium bromide (0.5 µg/mL) in TAE 1X. Expected size: bcl-2 MBR – 130bp +/ – 50–80 bp (*see* **Notes 11** and **12**).

3.3.1.2. Bcl-2 mcr Nested PCR

First PCR amplification:

1. Reaction mix (× *n* samples + 1) (*see* **Note 6**): dNTPs 2 m*M* 5 µL; 10X Buffer Taq polymerase (Promega, Madison, WI) 5 µL; 20 m*M* $MgCl_2$ 5 µL; mcr-2 (20 pmol/µL) 1 µL; JH3 (20 pmol/µL) 1 µl; Taq polymerase (Promega, Madison, WI) 0.2 µL (*see* **Note 7**); sterile water for a final volume of 50 µL.
2. Aliquot PCR reaction mix in each tube (*see* **Note 8**).
3. Add 0.5–1 µg of DNA sample in each tube except the no template control sample.
4. Amplification profile: first step, 94°C 3 min, 58°C 1 min, 72°C 1 min; second step × 30 cyles, 94°C 1 min, 58°C 1 min, 72°C 1 min ; third step, 72°C 10 min (*see* **Note 9**).

Second PCR amplification:

1. Reaction mix (× *n* samples + 1) (*see* **Note 6**): dNTPs 2 m*M* 5 µL; 10X Buffer Taq polymerase (Promega, Madison, WI) 5 µL; 20 m*M* $MgCl_2$ 5 µL; mcr-3.1 (20 pmol/µL) 1 µL; JH4 (20 pmol/µL) 1 µL; Taq DNA Polymerase (Promega, Madison, WI) 0.2 µL; sterile water for a final volume of 50 µL.
2. Aliquot PCR reaction mix in each tube.

3. Add 5 µL of first PCR product (*see* **Note 10**).
4. Amplification profile: first step, 94°C 3 min, 60°C 1 min, 72°C 1 min; second step × 30 cycles, 94°C 1 min, 60°C 1 min, 72°C 1 min; third step, 72°C 10 min (*see* **Note 9**).
5. Run 10 µL of PCR product (with 1 µL of loading buffer) and 10 µL of molecular weight marker (with 1 µL of loading buffer) on a 2% agarose gel stained with ethidium bromide (0.5 µg/mL) in TAE 1X. Expected size: bcl-2 mcr→ 550–600 bp ± 50–80 bp (*see* **Notes 11** and **12**).

3.3.2. Bcl-1/IgH Rearrangement

Mantle cell lymphoma is characterized by the t(11;14)(q13;q32) translocation, in which the bcl-1 gene on chromosome 11 is juxtaposed to the IgH locus on chromosome 14; the t(11;14) is detected by cytogenetic or Southern blot analysis in the vast majority of MCL patients (70% to 100%) ***(24)***. Breakpoints on chromosome 11 have been found dispersed over more than 100 kb of genomic DNA; the major translocation cluster (MTC) contains approx 30–50% of these translocations and it is a suitable target for PCR amplification. Because of this large breakpoint area less than 30% of MCL patients have a rearrangement that can be detectable by PCR.

The PCR reactions described below are used to identify the molecular marker as well as for MRD analysis (*see* **Note 6**). They must include:

- A bcl-1/IgH MTC positive sample to assess the sensitivity;
- Healthy donor polyclonal DNA as a negative sample to assess the specificity;
- A no-template control sample to exclude any possible contamination.

3.3.2.1. Bcl-1/IgH MTC Nested PCR

First PCR amplification.

1. Reaction mix (× *n* samples + 1) (*see* **Note 6**): dNTPs 2 m*M* 5 µL; 10X Buffer Taq polymerase (Applied Biosystems, Warrington, UK) 5 µL; 15 m*M* $MgCl_2$ 5 µL; P2 (10 pmol/µL) 1 µL; JH3 (10 pmol/µL) 1 µL; Taq polymerase (Applied Biosystems, Warrington, UK) 0.2 µL (*see* **Note 7**); sterile water for a final volume of 50 µL.
2. Aliquot PCR reaction mix in each tube (*see* **Note 8**).
3. Add 0.5–1 µg of DNA sample in each tube out of no-template control sample.
4. Amplification profile: first step × 33 cycles, 94°C 1 min, 58°C 30 s, 72°C 30 s; second step, 72°C 7 min (*see* **Note 9**).

Second PCR amplification:

1. Reaction mix (× *n* samples + 1) (*see* **Note 6**): dNTPs 2 m*M* 5 µL; 10X Buffer Taq polymerase (Applied Biosystems, Warrington, UK) 5 µL; 15 m*M* $MgCl_2$ 5 µL; P4 (10 pmol/µL) 1 µL; JH3 (10 pmol/µL) 1 µL; Taq DNA Polymerase (Applied Biosystems, Warrington UK) 0.2 µL; sterile water for a final volume of 50 µL.
2. Aliquot PCR reaction mix in each tube.
3. Add 2 µL of first PCR product (*see* **Note 10**).

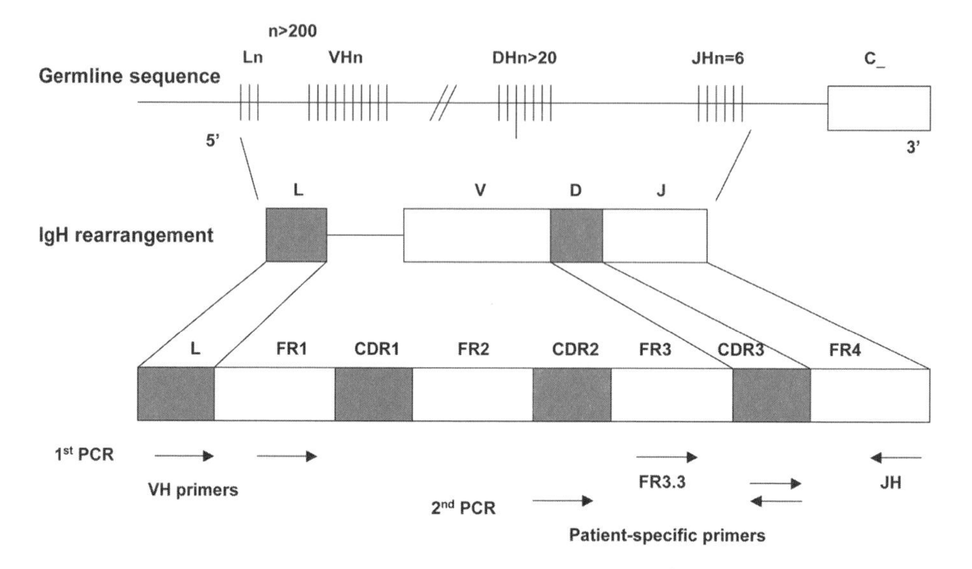

Fig. 1. Immunoglobulin heavy chain (IgH) gene rearrangement from the germline sequence. IgH gene leader (L), variable (V), diversity (D), and joining (J) region recombination is schematically represented. Arrows indicate the position of primer pairs used for PCR amplification of VDJ segment (first PCR) and for patient-specific nested or seminested PCR (second PCR).

4. Amplification profile: first step × 30 cycles, 94°C 1 min, 58°C 30 s, 72°C 30 s; second step: 72°C 7 min (*see* **Note 9**).
5. Run 10 µL of PCR product (with 1 µL of loading buffer) and 10 µL of molecular weight marker (with 1 µL of loading buffer) on a 2% agarose gel stained with ethidium bromide (0.5 µg/mL) in TAE 1X. Expected size: 250–350 bp (*see* **Note 11**).

3.3.3. VDJ Rearrangement

IgH rearrangement is an early event in B-lymphocyte ontogenesis and can be used as a molecular marker of the tumor clone in the vast majority of B-cell malignancies (**Fig. 1**). Three gene segments are present in the rearranged IgH genes namely the variable (V), diversity (D), and joining (J) regions. During B-lymphoid differentiation, the D segment joins J segment and the resulting D–J segment joins one V region sequence producing a VDJ complex ***(25)***. The enzyme terminal deoxynucleotidyltransferase (TdT) inserts random nucleotides at both the V–D and the D–J junctions and further diversity is generated by random excision of nucleotides and somatic mutations. The most hypervariable region produced by this phenomenon is the third complementaritydeterming region (CDRIII), which can be considered a unique marker for a given B-cell clone ***(26)***. The CDRIII region can be PCR-amplified using a variety of consensus framework region (FR) primers from VH and JH regions and it can be detected in about 80–90% of B-cell tumors ***(27)***. In order to have an adequate sensitivity and the specificity, the PCR product must be sequenced and patient-specific primers constructed within the CDR regions. These primers can be used for nested-PCR amplification of the tumor specific IgH rearrangement, thus allowing MRD detection with high sensitivity and specificity ***(28)***.

3.3.3.1. Identification of the VDJ Rearrangement with FR1 and Leader Forward Primers

The PCR reaction must include a no-template control sample with 1 μL of each VH primer to exclude any possible contamination.

1. Reaction mix (× *n* samples + 1) (*see* **Note 6**): dNTPs 2 m*M* 5 μL; 10X Buffer Taq polymerase (Promega, Madison, WI) 5 μL; 20 m*M* $MgCl_2$ 5 μL; JHD (20 pmol/μL) 1 μL; Taq polymerase (Promega, Madison, WI) 0.2 μL (*see* **Note 7**); add sterile water for a final volume of 50 μL.
2. Add 1 μL of one VH primer (20 pmol/μL) in each tube.
3. Aliquot PCR reaction mix in each tube.
4. Add 0.5–1 μg of DNA sample in each tube except the no-template control sample.
5. Amplification profile: first step, 94°C 1 min; second step × 35 cycles, 94°C 30 s, 62°C 30 s, 72°C 30 s; third step: 72°C 7 min (*see* **Note 9**).
6. Run 10 μL of PCR product (with 1 μL of loading buffer) and 10 μL of molecular weight marker (with 1 μL of loading buffer) on a 2% agarose gel stained with ethidium bromide (0.5 μg/mL) in TAE 1X. Expected size: VH/FS and VH/D→ 350–400 bp. VH/LS and VH/L → 500-550 bp (*see* **Note 13**).

3.3.3.2. Identification of the VDJ Rearrangement with FR3-Derived Forward Primer

The PCR reaction must include a no template control sample to exclude any possible contamination.

1. Reaction mix (× *n* samples + 1) (*see* **Note 6**): dNTPs 2 m*M* 5 μL; 10X Buffer Taq polymerase (Promega, Madison, WI) 5 μL; 20 m*M*$MgCl_2$ 5 μL; FR3.3 (20 pmol/μL) 1 μL; JH3.3 (20 pmol/μL) 1 μL; Taq polymerase (Promega, Madison, WI) 0.2 μL (*see* **Note 7**); sterile water to a final volume of 50 μL.
2. Aliquot PCR reaction mix in each tube.
3. Add 0.5–1 μg of DNA sample in each tube out of no-template control sample.
4. Amplification profile: first step, 94°C 1 min; second step × 35 cycles, 94°C 30 s, 55°C 30 s, 72°C 30 s; third step, 72°C 7 min (*see* **Note 9**).
5. Run 5 μL of PCR product (with 1 μL of loading buffer) and 5 μL of molecular weight marker (with 1μL of loading buffer) on 12% polyacrylamide gel (19:1 acrylamide: bisacrylamide) in 1X TAE buffer. Expected size: ≤120 bp.

3.3.3.3. PCR Product Sequencing and Internal Primers Design

Clonal amplifications give sharp band of the appropriate size; these PCR products can be electrophoresed through a 0.8% LMP agarose gel, excised, and purified using QIAquick gel extraction kits (Qiagen, Milan, Italy) (*see* **Note 14**). Direct sequencing of amplified DNAs can be perfomed using the Promega sequencing system as previously described ***(29)*** or automated sequencing.

Confirm that the sequence derives from IgH rearrangement. Exclusion of IgH sequences bearing stop codons and appropriate identification of germline V, D, and J regions can be performed at www.ncbi.nlm.nih.gov/igblast/ or http://imgt.cines.fr/home.html; hypervariable regions CDRII and CDRIII should be identified as the junctions including N-insert (*see* **Note 15**), from which patient-specific oligonucleotide primers can be designed. If FR3-derived forward primer has been used, the length of the DNA sequence allows only the design of a patient-specific forward primer within the CDRIII region for a seminested PCR to be performed with JH-derived consensus primer as reverse primer **(Fig. 1)**.

These allele-specific oligonucleotides (ASO) primers are usually 18–20 bp long with a melting temperature of approx 60°C (*see* **Note 16**). In order to determine the optimal PCR conditions, a so called "Oligotest" should be performed to test the PCR amplification of a diagnostic sample with different primer and $MgCl_2$ concentrations (usually 1 µL of 10–20 pmol/µL primers and 5 µL of 15–25 m*M* $MgCl_2$, respectively), annealing temperature (usually the mean temperature of the ASO primers), and second-step cycle number (usually between 30 and 35 cycles).

3.3.3.4. Minimal Residual Disease Monitoring with IgH Molecular Marker

The PCR reaction must include:

- A patient positive sample to assess the sensitivity;
- Healthy donor polyclonal DNA as a negative sample to assess the specificity;
- A no-template control sample to exclude any possible contamination.

First PCR amplification:

Performed as described under **Subheading 3.3.3.1.** or **3.3.3.2.**, using only the patient IgH forward primer.

Second PCR amplification:

1. Reaction mix (× n samples + 1) (*see* **Note 6**): dNTPs 2 m*M* 5 µL; 10X Buffer TaqGold polymerase (Applied Biosystems, Warrington, UK) 5 µL; $MgCl_2$ (as defined by "Oligotest") 5 µL; AmpliTaq Gold (Applied Biosystems, Warrington, UK) 0.2 µL; patient-specific forward primer (10–20 pmol/µL or as defined by "Oligotest") 1 µL; patient-specific reverse primer (10–20 pmol/µL or as defined by "Oligotest") 1 µL; sterile water for a final volume of 50 µL.

2. Aliquot PCR reaction mix in each tube.
3. Add 1 μL of first PCR product (*see* **Note 10**).
4. Amplification profile: as defined by "Oligotest" (*see* **Note 9**).
5. Run 10 μL of PCR product (with 1 μL of loading buffer) and 10 μL of molecular weight marker (with 1 μL of loading buffer) on a 2% agarose gel stained with ethidium bromide (0.5 μg/mL) in TAE 1X (*see* **Note 11**). If the expected size is below 100–120 bp, run 5 μL of PCR product and 5 μL of molecular weight marker on 12% polyacrilamide gel (19:1 acrylamide: bisacrylamide) in TAE 1X.

2.4 Real-time Quantitative PCR

Although different equipments and techniques can be utilized for real-time PCR-based quantitative MRD monitoring *(30)*, here we present in detail the more common strategy. This strategy utilizes a hydrolysis probe approach on a high-throughput instrument (ABI Prism 7000 Sequence Detection System Instrument, Applied Biosystems, Warrington, UK) for absolute quantification of the tumor genome number.

The hydrolysis probe approach is based on the 5′ exo-nuclease activity of the Taq polymerase and requires the synthesis of three oligonucleotides: one forward primer, one reverse primer, and one probe. The probe is labeled at the 5′ end with reporter fluorochrome (e.g., 6-carboxyfluorescein [6-FAM]) and at the 3′ end with a quencher fluorochrome (e.g., 6-carboxy-tetramethyl-rhodamine [TAMRA]). This oligonucleotide is designed to be inside the amplicon between the two primers in order to be cleaved by the Taq polymerase 5′ exo-nuclease activity during the amplification. As a result of this cleavage, reporter fluorochrome is separated from quencher and the fluorescence can be detected by the instrument. Specific PCR products quantity can thus be detected as the reaction proceeds. In lymphomas with chromosomal translocations, consensus primers and probes suitable for a large number of patients can be used to target the specific translocation *(31)*. A Taqman system (primer pair and fluorogenic probe) for quantification of the bcl-2-MBR rearrangement is detailed in Table 3.

When the molecular marker is based on the IgH gene rearrangement, the target gene is unique for each patient, thus the Taqman system should be designed on each IgH sequence. For this purpose, different strategies have been proposed. In the "ASO probe approach," the probe is positioned in the CDRIII region of the IgH gene rearrangement, while the forward and reverse primer are designed in VH and JH germline sequences, respectively. Because patient-specific probes are expensive and time-consuming, assays based on a limited number of probes designed within the FR3 region have been developed *(32)*. In this case, the specificity is given by forward and reverse primers, which are located in the VH and CDRIII segments, respectively. As an alternative, the forward primer can be positioned in the CDRIII region and the reverse primer and probe within the JH segment. For primer and probe design, several software packages are available (e.g., Primer Express®, Applied Biosystems, Warrington, UK). When a consensus probe positioned in the FR3 region of the IgH gene is used, the patient-specific primers designed for qualitative nested PCR can also be suitable for real-time PCR. However, the following parameters should always be satisfied: probe melting temperature 10°C higher than the primers, primers melting temperature between 58 and 60°C, amplicon size between 50 and 150 bp, no 5′ G residue, and no more than three consecutive G residues.

In order to normalize patient samples for DNA quality and quantity, a control gene must be included in every real-time PCR analysis. The glyceraldehyde-3-phosphate dehydrogenase (GAPDH) is often used as reference standard (*see* **Table 3**), but other genes, such as β- actin, β-globin, Abelson, RNase P, and albumin genes, can be chosen as well.

In order to quantify the target and reference genes, standard curves are generated by serial dilutions of plasmids containing the DNA target sequence. If 10-fold dilutions are used, the slope of a standard curve should be –3.3 (–3.0– –3.9) with a correlation coefficient >0.98. PCR products are cloned by means of a TA cloning kit (*see* **Subheading 3.3.3.3.**); GAPDH is cloned from normal peripheral blood mononuclear cells, bcl-2/MBR gene from positive cell lines (e.g., DOHH-2), and IgH patient-specific rearrangement from a diagnostic sample.

Each real-time PCR plate must include:

- Samples with serial 10-fold dilution of bcl-2 or IgH plasmid into a pool of DNA from 5–10 samples of normal peripheral blood mononuclear cells and reference gene into sterile water (plasmid copy number ranging from 10^{-6} to 10^{-1}) to produce the standard curve;
- Healthy donor polyclonal DNA as a negative sample to assess the specificity;
- A no-template control sample to exclude any possible contamination.

1. Reaction mix (× *n* samples + 1): TaqMan Universal PCR Master Mix (Applied Biosystems, Warrington, UK) 25 μL; 1 μL of each primer (20 pmol/μL), 1 μL of probe (10 pmol/μL); sterile water to a final volume of 50 μL (*see* **Notes 6** and **17**).
2. Aliquot PCR reaction mix in each tube.
3. Add 0.5–0.6 μg in 2.5 μL of DNA sample in each tube except the no-template control sample.
4. Amplification profile: first step, 50°C 2 min; second step, 95°C 10 min; third step (for 45 cycles), 95°C 15 s, 60°C 1 min (*see* **Note 8 and 18**).

   In each plate, all the samples are run in triplicate. The MRD value is calculated as the mean of triplicates of target gene divided by the mean of triplicates of GAPDH as control gene.

4. Notes

1. Filter dense samples using a 5-mL sterile syringe.
2. If PBS is not available, 0.9% sodium chloride solution is also suitable.

3. For DNA extraction, other methods can be used, such as phenol/chloroform DNA precipitation, which provides the same DNA quality and quantity but it is more time-consuming. DNA precipitation by means of DNA binding columns (e.g., Nucleospin®, BD Biosciences, France) is particularly useful when less than 5 × 10^6 cells are available.
4. Use the following protocol for paraffin-embedded tissue: add 1 mL of xylol to two sections of paraffin-embedded tissue and incubate at 37°C in water bath for 15 min; spin for 3 min at 13000 rpm; discard the supernatant and resuspend the pellet with 1 mL of xylol, repeating all the steps described above; then add 1 mL of 100% ethanol an mix; incubate 5 min at room temperature; spin at 13000 rpm for 3 min and discard the supernatant; add 1 mL of 100% ethanol and repeat the step described above; dry the pellet using a Speed-Vac; resuspend the dried pellet with TNE 200 μL, sodium dodecyl sulphate (SDS) 20% 10 μL and proteinase K (25 mg/mL) 4 μL and incubate at 55°C overnight, adding 2 μL of proteinase K after 2–3 h; extract once with one volume of phenol and mix gently; spin for 10 min at 13000 rpm; transfer the upper phase and extract once with phenol/sevag; spin for 10 min at 13,000 rpm; recover the upper phase and extract once with 1 volume of Sevag; spin for 10 min at 13,000 rpm; transfer the upper phase and precipitate with 2 volumes of 100% ethanol and one-tenth volume of 3 *M* Na-Acetate pH 8.2; spin for 10 min at 13,000 rpm; wash the pellet with 70% ethanol and spin for 10 min at 13,000 rpm; dry the pellet at room temperature and resuspend in 8 m*M* NaOH.
5. If the pellet is very small, spin 5 min at 5000 rpm, discard the supernatant, and dry at room temperature.
6. The PCR mix is set up in the "PCR area" and, in order to avoid contamination, DNA is added on a different bench and with dedicated instruments (always use sterilized aerosol-resistant tips); PCR machines and PCR products are located in the "post-PCR area"; for nested PCR, first PCR product amplification is added in this room.
7. PCR amplifications that use Taq polymerase Promega must be set up on ice.
8. All the patient samples that are available for MRD studies are amplified in triplicate; this allows the detection of tumor cells that are just at the limit of assay sensitivity.
9. PCR products can be kept on ice until they are in the thermal cycler and then stored at –20°C in the post-PCR area.
10. To avoid contamination during the nested PCR reactions, use dedicate instruments (pipet and tips); keep each tube capped except when adding each respective first PCR product; start adding PCR product from negative samples, patient sample, and, at the end, positive control sample (it is advisable to respect this order in all the PCR amplifications).
11. In order to avoid false-negative results, all of the MRD-negative samples must be tested with PCR amplification of a gene located on a chromosome not frequently involved in the malignancy, e.g., β-actin.
12. If the band size is different, PCR product must be sequenced.

13. In some cases, if any sharp band can be seen, it could be useful to repeat the experiment using total cDNA.
14. If no other band can be seen over and under the clonal band, the PCR product can be purified with QIAquick PCR purification kit (QIAGEN, Milan, Italy).
15. When the sequence quality does not allow a complete reading of CDR regions, the PCR product can be cloned with TA cloning kit. Plasmid DNA is isolated by miniprep and, after restriction enzyme analysis, those containing the insert are sequenced (the IgH rearrangement of the tumor clone is found if at least 60% of 15–20 sequenced plasmid DNAs have the same insert).
16. Avoid primers that can fold and anneal against themselves or with other copies of themselves or with the reverse primer; avoid more than three adjacent G or C.
17. In order to inhibit PCR inhibitors in a patient sample, 0.5 μL/tube of serum bovine albumin can be added to the PCR reaction mix.
18. The third step can be modified in order to improve primers annealing and reaction efficiency.

References

1. Corradini, P., Ladetto, M., Zallio, F. et al. (2004) Long-term follow-up of indolent lymphoma patients treated with high-dose sequential chemotherapy and autografting: evidence that durable molecular and clinical remission frequently can be attained only in follicular subtypes. *J. Clin. Oncol.* **22**, 1460–1468.
2. Hunault-Berger, M., Ifrah, N., Solal-Celigny, P., Groupe Ouest-Est des Leucemies Aiguee et des Maladies du Sang (2002) Intensive therapies in follicular non-Hodgkin lymphomas. *Blood* **100**, 1141–1152.
3. Gribben, J. G., Freedman, A., Woo, S. D., et al. (1991) All advanced stage non-Hodgkin's lymphomas with a polymerase chain reaction amplifiable breakpoint of bcl-2 have residual cells containing the bcl-2 rearrangement at evaluation and after treatment. *Blood* **78**, 3275–3280.
4. Gribben, J. G., Freedman, A. S., Neuberg, D., et al. (1991) Immunologic purging of marrow assessed by PCR before autologous bone marrow transplantation for B-cell lymphoma. *N. Eng. J. Med.* **325**, 1525–1533.
5. Gribben, J. G., Neuberg, D., Freedman, A. S., et al. (1993) Detection by polymerase chain reaction of residual cells with the bcl-2 traslocation is associated with increased risk for relapse after autologous bone marrow transplantation for B-cell lymphoma. *Blood* **81**, 3449–3457.
6. Freedman, A. S., Neuberg, D., Mauch, P., et al. (1999) Long-term follow up of autologous bone marrow transplantation in patients with relapsed follicular lymphoma. *Blood* **94,** 3325–3333.
7. Moos, M., Schulz, R., Martin, S., et al (1998) The remission status before and the PCR status after high-dose therapy with peripheral blood stem cell support are prognostic factors for relapse-free survival in patients with follicular non-Hodgkin's lymphoma. *Leukemia* **12**, 1971–1976.
8. Gianni, A. M., Magni, M., Martelli, M., et al. (2003) Long-term remission in mantle cell lymphoma following high-dose sequential chemotherapy and in vivo

rituximab-purged stem cell autografting (R-HDS regimen). *Blood* **102**, 749–755.

9. van Besien, K., Keralavarma, B., Devine, S., and Stock, W. (2001) Allogeneic and autologous transplantation for chronic lymphocytic leukemia. *Leukemia* **15**, 1317–1325.
10. Dreger, P. and Montserrat, E. (2002) Autologous and allogeneic stem cell transplantation for chronic lymphocytic leukemia. *Leukemia* **16**, 985–992.
11. Milligan, D. W., Fernandes, S., Dasgupta, R., et al (2005) Results of the MRC pilot study show autografting for younger patients with chronic lymphocytic leukemia is safe and achieves a high percentage of molecular responses. *Blood* **105**, 397–404.
12. Pavletic, Z. S., Bierman, P. J., Vose, J. M., et al. (1998) High incidence of relapse after autologous stem-cell transplantation for B-cell chronic lymphocytic leukemia or small lymphocytic lymphoma. *Ann. Oncol.* **9**, 1023–1026.
13. Esteve, J., Villamor, N., Colomer, D., et al. (2001) Stem cell transplantation for chronic lymphocytic leukemia: different outcome after autologous and allogeneic transplantation and correlation with minimal residual disease status. *Leukemia* **15**, 445–451.
14. Provan, D., Bartlett-Pandite, L., Zwicky, C., et al. (1996) Eradication of polymerase chain reaction-detectable chronic lymphocytic leukemia cells is associated with improved outcome after bone marrow transplantation. *Blood* **88**, 2228–2235.
15. Szczepanski, T., Willemse, M. J., Brinkhof, B., et al. (2002) Comparative analysis of Ig and TCR gene rearrangements at diagnosis and at relapse of childhood precursor-B-ALL provides improved strategies for selection of stable PCR targets for monitoring of minimal residual disease. *Blood* **99**, 2315–2323.
16. van Dongen, J. J., Seriu, T., Panzer-Grumayer, E. R., et al. (1998) Prognostic value of minimal residual disease in acute lymphoblastic leukaemia in childhood. *Lancet* **352**, 1731–1738.
17. Cave, H., van der Werff ten Bosch, J., Suciu, S., et al. (1998) Clinical significance of minimal residual disease in childhood acute lymphoblastic leukemia. European Organization for Research and Treatment of Cancer—Childhood Leukemia Cooperative Group. *N. Eng. J. Med.* **339,** 591--598.
18. Mortuza, F. Y., Papaioannou, M., Moreira, I. M., et al. (2002) Minimal residual disease tests provide an independent predictor of clinical outcome in adult acute lymphoblastic leukemia. *J. Clin. Oncol.* **20**, 1094–1104.
19. Holland, P. M., Abramson, R. D., Watson, R., and Gelfand, D. H. (1991) Detection of specific polymerase chain reaction product by utilizing the 5′→3′ exonuclease activity of Thermus aquaticus DNA polymerase. *Proc Natl Acad Sci USA* **88**, 7276–7280.
20. Ladetto, M., Sametti, S., Donovan, J. W., et al. (2001) A validated real-time quantitative PCR approach shows a correlation berween tumor burden and successful ex vivo purging in follicular lymphoma patients. *J. Exp. Hematol.* **29**, 183–193.

21. Tsujimoto, Y., Finger, L. R., Yunis, J., et al. (1984) Cloning of the chromosome breakpoint of neoplastic B cells with the t(14;18) chromosome translocation. *Science* **226**, 1097–1099.
22. Cleary, M. L. and Sklar, J. (1985) Nucleotide sequence of a t(14;18) chromosomal breakpoint in follicular lymphoma and demonstration of a breakpoint-cluster region near a transcriptionally active locus on chromosome 18. *Proc Natl Acad Sci USA* **82**, 7439–7443.
23. Cleary, M. L., Galili, N., and Sklar, J. (1986) Detection of a sond t(14;18) breakpoint cluster region in human follicular lymphomas. *J. Exp. Med.* **164**, 315–320.
24. Tsujimoto, Y., Yunis, J., Onorato-Showe, L., et al. (1984) Molecular cloning of the chromosomal breakpoint of B-cell lymphomas and leukemias with the t(11;14) chromosome traslocation. *Science* **224**, 1403–1406.
25. Early, P., Huang, H., Davis, M., et al. (1980) An immunoglobulin heavy chain variable region gene is generated from three segments of DNA. *Cell* **19**, 281.
26. Tonegawa, S. (1983) Somatic generation of antibody diversity. *Nature* **302**, 575–581.
27. Deane, M., McCarthy, K. P., Wiedemann, L. M., and Norton, J. D. (1991) An improved method for detection of B-lymphoid clonality by polymerase chain reaction. *Leukemia* **5,** 726–730.
28. Voena, C., Ladetto, M., Astolfi, M., et al. (1997) A novel nested-PCR strategy for detection of rearranged immunoglobulin heavy chain genes in B-cell tumours. *Leukemia* **11**, 1793–1798.
29. Corradini, P., Ladetto, M., Voena, C., et al. (1993) Mutational activation of N- and K-ras oncogenes in plasma cell dyscrasias. *Blood* **81**, 2708–2713.
30. Van der Velden, V. H. J., Hochhaus, A., Cazzaniga, G., Szczepanski, T., Gabert, J., and van Dongen, J. J. M. (2003) Detection of minimal residual disease in hematologic malignancies by real-time quantitative PCR: principles, approaches, and laboratory aspects. *Leukemia* **17,** 1013–1034
31. Luthra, R., McBride, J. A., Cabanillas, F., and Sarris, A. (1998) Novel 5′ exonuclease-based real-time PCR assay for the detection of t(14;18)(q32;q21)in patients with follicular lymphoma. *Am. J. Pathol.* **153**, 63–68.
32. Donovan, J. W., Ladetto, M., Zou, G., et al. (2000) Immunoglobulin heavy-chain consensus probes for real-time PCR quantification of residual disease in acute lymphoblastic leukemia. *Blood* **95**, 2651–2658.

14

Detection of Impending Graft Rejection and Relapse by Lineage-Specific Chimerism Analysis

Thomas Lion

Summary

Molecular surveillance of hematopoietic chimerism has become part of the routine diagnostic program in patients after allogeneic stem cell transplantation. Chimerism testing permits early prediction and documentation of successful engraftment, and facilitates early detection of impending graft rejection. In patients transplanted for treatment of malignant hematological disorders, monitoring of chimerism can provide an early indication of incipient disease relapse. The investigation of chimerism has therefore become an indispensable tool for the management of patients during the posttransplant period. Growing use of nonmyeloablative conditioning, which is associated with prolonged duration of mixed hematopoietic chimerism, has further increased the clinical importance of chimerism analysis. At present, the most commonly used technical approach to the investigation of chimerism is microsatellite analysis by PCR. The investigation of chimerism within specific leukocyte subsets isolated from peripheral blood or bone marrow samples by flow-sorting or magnetic beads-based techniques provides more specific information on processes underlying the dynamics of donor/recipient chimerism. Moreover, cell subset-specific analysis permits the assessment of impending complications at a significantly higher sensitivity, thus providing a basis for earlier treatment decisions.

Key Words: Chimerism; microsatellites; leukocyte subsets; flow-sorting; graft rejection; relapse.

1. Introduction

Investigation of donor- and recipient-derived hemopoiesis (chimerism) by molecular techniques facilitates the monitoring of engraftment kinetics in patients after allogeneic stem cell transplantation. The analysis of chimerism during the immediate posttransplant period permits early assessment of successful engraftment or graft failure *(1,2)*. Patients receiving reduced intensity (nonmyeloablative) conditioning regimens have an increased risk of graft rejection, particularly

From: *Methods in Molecular Medicine, vol. 134: Bone Marrow and Stem Cell Transplantation*
Edited by: M. Beksac © Humana Press Inc., Totowa, NJ

if they are transplanted with T-cell-depleted grafts *(3)*. The monitoring of residual recipient natural killer (NK)- and T-cells in peripheral blood (PB) can provide timely indication of impending allograft rejection *(3,4)*. In patients undergoing allogeneic stem cell transplantation for the treatment of leukemia, impending disease recurrence can be indicated by an increasing proportion of recipient-derived cells *(4,5)*. PCR-based chimerism assays analyzing highly polymorphic microsatellite (short tandem repeat [STR]) markers permit the detection of residual autologous cells at a sensitivity of about 1–5% *(6–11)*. When investigating chimerism in total leukocyte preparations from PB, this level of sensitivity may not be sufficient to allow early assessment of impending complications. It is possible to overcome this problem by investigating chimerism in specific leukocyte subsets of interest isolated by flow-sorting or by immunomagnetic bead separation. Because residual recipient-derived cells can be detected within the individual leukocyte fractions with similar sensitivity, it is possible to identify and monitor minor autologous populations that escape detection in total PB leukocyte samples. The overall sensitivity of chimerism assays achievable by investigating specifically enriched leukocyte subsets is in a range of 0.1–0.01% *(4)*, i.e., one to two logs higher than analysis of total leukocyte preparations.

1.1. Prediction of Graft Rejection by the Monitoring of Chimerism Within Lymphocyte Subsets

Patients who receive reduced intensity conditioning reveal persisting leukocytes of recipient genotype more commonly than patients after myeloablative conditioning. The higher incidence of mixed or recipient chimerism may be attributable both to cells of myeloid and lymphoid lineages *(3)*. In patients who receive T-cell-depleted grafts, there is a strong correlation with the presence of mixed or recipient chimerism within T-cells (CD3+) and NK-cells (CD56+). Detection of mixed chimerism within lymphocyte populations is associated with an increased risk of late rejection *(3,4)*. In most instances, serial analysis reveals a persistently high or an increasing recipient-specific allelic pattern prior to overt graft rejection *(3–5)* (**Fig. 1**). The correlation between the observation of mixed or recipient chimerism and graft rejection was shown to be higher for NK-cells than for T-helper (CD3+/CD4+) or T-suppressor (CD3+/CD8+) cells *(3)*. Patients displaying recipient chimerism in CD56+ cells between days +14 and +35 appear to have an extremely high risk of graft rejection. By contrast, virtually all patients who experience late graft rejection show pure donor genotype within the myeloid (CD14+ and CD15+) cells during the same period *(3)*. Hence, the observation of recipient chimerism within the CD56+ and CD3+ cell subsets is highly predictive for the occurrence of late graft rejection. These findings underscore the importance of cell subset analysis in post-transplant chimerism testing.

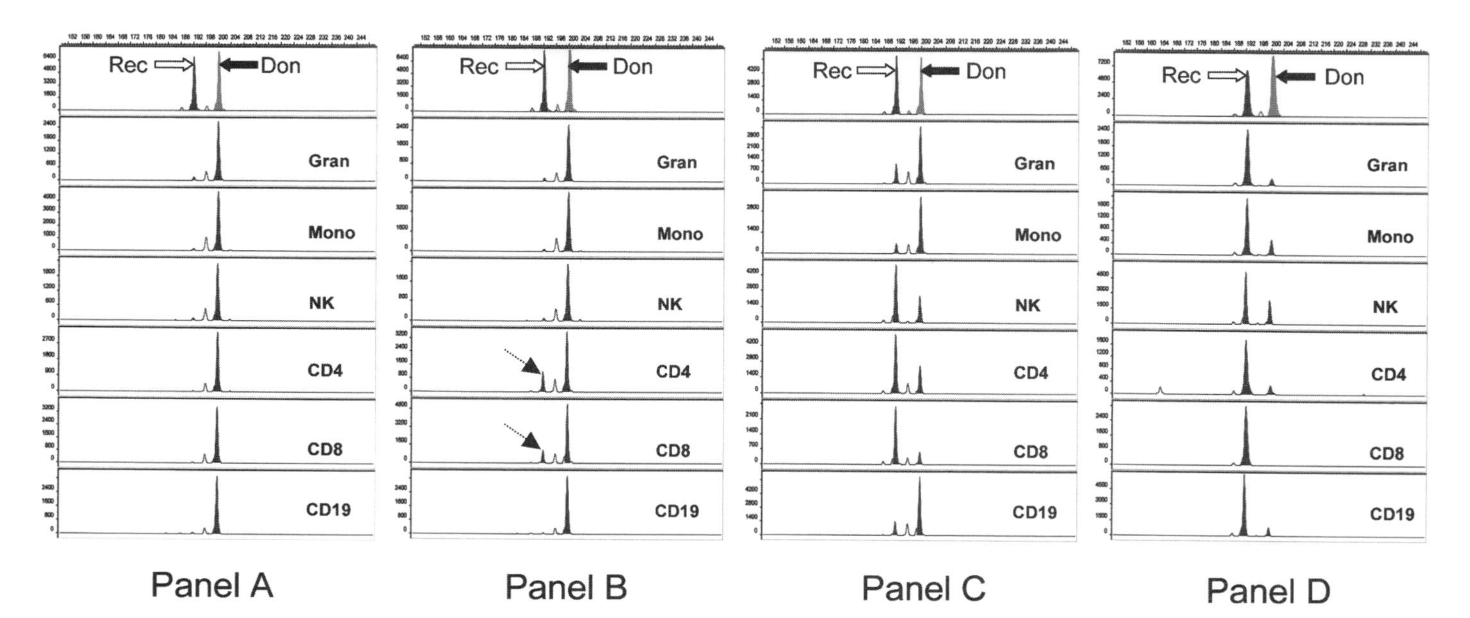

Fig. 1. Detection of impending graft rejection by lineage-specific chimerism analysis. **A–D** represent sequential time points of lineage-specific chimerism testing in a patient who ultimately rejected the allograft. **(A)** Reveals the presence of virtually pure donor chimerism in all leukocyte subsets analyzed, at the level of sensitivity achievable by the technique used (*see* **Note 12**). **(B)** Mixed chimerism appears in CD4 and CD8 lymphocyte subsets (dotted arrows), while donor chimerism is present in all other populations tested. **(C)** Reveals the predominance of autologous cells within the CD4, CD8, and NK cell subsets, and all other fractions show a small proportion of recipient-derived cells. This constellation is highly suggestive of imminent graft rejection. **(D)** Documents autologous recovery in all cell populations, with minor proportions of residual donor cells in some fractions.

Don, donor-specific allele, indicated by black arrow; Rec, recipient-specific allele, indicated by white arrow (in this example, both donor and recipient are homozygous for the marker selected, and thus display only one allele, respectively); Gran, granulocytes; Mono, monocytes; NK, natural killer cells; CD4, helper T-lymphocytes; CD8, suppressor T-lymphocytes; CD19, B-lymphocytes.

1.2. Detection of Imminent Relapse by Serial Analysis of Leukemia Lineage-Specific Chimerism

Investigation of entire leukocyte fractions from PB has been shown to reveal reappearance of autologous cells (mixed chimerism) before the diagnosis of relapse *(5)*, thus providing a basis for timely initiation of treatment, which usually includes the reduction or withdrawal of immunosuppressive therapy or the administration donor lymphocyte infusions (DLI). In some instances, however, analysis of chimerism within total leukocytes may not shown any changes indicative of impending relapse *(1,4)*. Owing to its higher sensitivity, serial investigation of specific leukocyte subsets derived from PB or bone marrow (BM) has a greater potential of revealing informative changes in patients who later experience hematological relapse, provided that chimerism testing is performed at adequate intervals. These patients reveal either persistence or reappearance of autologous allelic patterns within cell populations expected to harbor leukemic cells, if present (**Fig. 2**). These cell populations can be specifically enriched for chimerism testing by targeting the original immunophenotype of the leukemic clone. The stem cell marker CD34 is commonly expressed by the leukemic cells in combination with lineage specific markers. For example, the leukemic clone in B-cell precursor acute lymphocytic leukemias is generally characterized by co-expression of CD34 and CD19. The cell populations expressing these markers can therefore be specifically targeted for the assessment of residual disease by monitoring the presence and the kinetics of recipient chimerism (**Fig. 2**). Occasionally, however, the only observation made before hematological relapse is lineage-specific chimerism kinetics suggestive of graft rejection *(4)*. This observation may be attributable to the loss of the graft-vs-leukemia effect associated with rejection of the allograft.

1.3. Technical Aspects

Despite the recent introduction of single-nucleotide polymorphism (SNP) and insertion/deletion (Indel) polymorphism analysis by real-time PCR for the investigation of chimerism *(12–14)*, STRs have remained the most commonly used source of polymorphic markers in the human genome for quantitative assessment of donor/recipient hemopoiesis after allogeneic stem cell transplantation. The exploitation of SNP and Indel markers for chimerism testing are promising approaches, which can provide a sensitivity superior to STR analysis by PCR. However, the spectrum of well characterized Indel polymorphisms must be expanded to provide an informative marker in any recipient/donor constellation, and currently available SNP-based techniques appear to lack the precision required for monitoring of chimerism kinetics in the range between 1 and 100% donor/recipient cells. Within this range, STR-PCR and, in the sex-mismatched transplant setting, fluorescence *in situ* hybridization (FISH) analy-

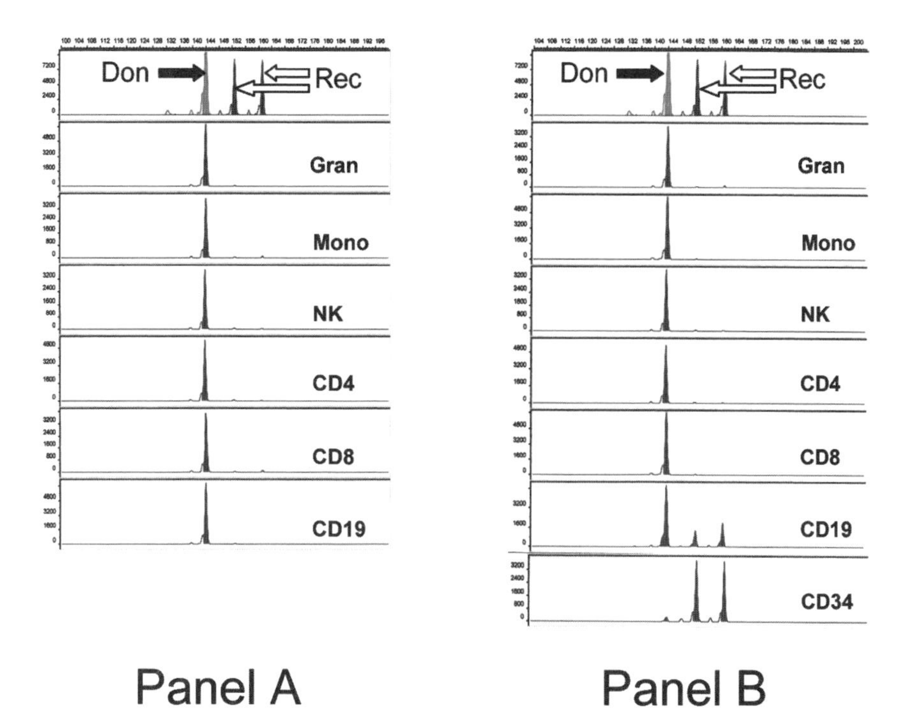

Fig. 2. Detection of incipient relapse by lineage-specific chimerism analysis. In this example, surveillance of residual disease after allogeneic stem cell transplantation is shown for a patient with B-cell precursor acute lymphocytic leukemia. The immunophenotype of the original leukemic clone revealed co-expression of CD34 and CD19. Residual or reappearing leukemic cells are therefore expected to occur within these leukocyte fractions. A and B show the investigation of different leukocyte subsets isolated by flow-sorting from peripheral blood at two sequential time points. A reveals the presence of complete donor chimerism in all leukocyte subsets analyzed, at the level of sensitivity achievable by the technique used (*see* **Note 12**). In B, the CD19-positive cell fraction shows mixed chimerism, indicating the reappearance of autologous cells. The CD34-positive cell population, which was not detectable at all at the previous time point of chimerism analysis (A), consists of recipient-derived cells. This finding is highly suggestive of reappearance of the leukemic clone. At this time point, the size of the leukemic clone may be well below the detection level of chimerism testing within total white blood cells. In instances in which a leukemia- or clone-specific marker is available (e.g., translocation-associated gene rearrangement, immunoglobulin- or T-cell receptor rearrangement), the finding can be controlled by an independent technique. In cases in which no other informative marker is available, lineage-specific chimerism testing is the most sensitive approach to early detection of impending relapse.

Don, donor-specific allele, indicated by black arrow; Rec, recipient-specific alleles, indicated by white arrow (in this example, the donor is homozygous for the marker selected, and thus displays only one allele, while the recipient is heterozygous); Gran, granulocytes; Mono, monocytes; NK, natural killer cells; CD4, helper T-lymphocytes; CD8, suppressor T-lymphocytes; CD19, B-lymphocytes; CD34, hematopoietic stem cells.

sis of the X and Y chromosomes ***(15)***, are presently the techniques providing the greatest accuracy in quantitative investigation of chimerism. As outlined previously, the overall sensitivity and the clinical utility of diagnostic information can be greatly increased by performing the analysis of chimerism in specific leukocyte fractions isolated from PB or BM. In this chapter, particular emphasis has therefore been put on the technical requirements of lineage-specific investigation of chimerism.

2. Materials

2.1. Isolation of Cell Subsets by Flow-Sorting

2.1.1. Hardware

1. White blood cell (WBC) counter.
2. Refrigerated centrifuge: Megafuge 1.0 (Heraeus).
3. Flow-sorter: FACS-Aria (Beckton Dickinson).
4. 50 mL polypropylene conical tubes (Beckton Dickinson).
5. 15 mL polypropylene conical tubes (Beckton Dickinson).
6. FACS tubes: 5 mL polystyrene tubes (Becton Dickinson).
7. Filters: 40 µm mesh (Nybolt).
8. Vortex mixer.

2.1.2. Reagents

1. Red blood cell (RBC) lysis buffer: 8.3 g NH_4Cl/1 g $KHCO_3$ per L; pH is adjusted to 7.4 with buffer (citrate-HCl pH = 4.0 or citrate-NaOH pH = 6.0 [Merck]).
2. Phosphate-buffered saline (PBS): NaCl 7.597 g/$Na_2HPO4 \times 2\ H_2O$ 1.245 g/$NaH_2PO_4 \times H_2O$ 0.414 g per L.
3. MOPC-21 (lyophilized reagent, Sigma M7894).
4. Fetal calf serum (FCS; Sebak).
5. Penicillin/streptomycin (PS; Gibco).
6. Na-Azide (Merck).
7. MOPC-21 solution: 5 mg lyophilized reagent/5 mL PBS containing 2% FCS)/PS 1:100 Vol/10% Na-Azide.
8. Monoclonal antibodies (MAbs):

Syto 41	Stains all nucleated cells (Eubio)
8FITC	Stains CD8- suppressor T-cell marker (Dako)
56PE	Stains CD56-NK-cell marker (Becton Dickinson)
3ECD	Stains CD3- T-cell marker (Instrument Laboratory)
45PerCP	Stains CD45-Pan-leukocyte marker (Becton Dickinson)
4PE-Cy7	Stains CD4- helper T-cell marker(Becton Dickinson)
71APC	Stains CD71-normocyte marker (Becton Dickinson)
14APC-Cy7	Stains CD14-monocyte marker (Becton Dickinson)
15FITC	Stains CD15-granulocyte marker (Dako)
33 PE	Stains CD33-myeloid cell marker (Becton Dickinson)

19ECD	Stains CD19-B lymphocyte marker (Instrument Laboratory)
34PE-Cy7	Stains CD34-stem cell marker (Becton Dickinson)

9. RPMI 1640 Medium without L-Glutamine (Gibco).

2.2. Purification of DNA

2.2.1. Purification of DNA From Peripheral Blood or Bone Marrow

2.2.1.1. Hardware

1. Spectrophototometer.
2. Benchtop centrifuge.
3. Heat block or water bath.
4. Vortex mixer.

2.2.1.2. Reagents

1. Qiagen DNA Blood Mini Kit (Qiagen, Hilden, Germany).
2. Ethanol (absolute).

2.2.2. Purification of DNA From Nails: Variant A, In-House Method

2.2.2.1. Hardware

1. Benchtop centrifuge.
2. Heat block or water bath.
3. Vortex mixer.

2.2.2.2. Reagents

1. Washing solution: 0.5% sodium dodecyl sulfate (SDS) and 0.5*M* NaOH.
2. dH_2O (DNAase free).
3. Extraction buffer: 1X TBE, 1.4% SDS, 0.14*M* NaCl, 0.28 *M* dithiothreitol (DTT) and 68 μg/mL proteinase K.
4. Phenol/chloroform /isoamylalcohol (Amresco, Solon, OH).
5. Chloroform/isoamylalcohol 25:1.
6. 3 *M* Na-Acetate, pH 5.2.
7. Ethanol (absolute).
8. Ethanol 70%.

2.2.3. Purification of DNA From Nails: Variant B, Tissue and Hair Extraction Kit with the DNA IQ™ System; Promega

2.2.3.1. Hardware

1. Benchtop centrifuge.
2. Heat block or water bath.
3. Vortex mixer.
4. MagneSphere® Technology Magnetic Separation Stand) Promega, Madison, WI).

2.2.3.2. Reagents

1. Tissue and Hair Extraction Kit (Promega, Madison,WI).
2. DNA IQ System (Promega, Madison,WI).
3. Nuclease-free water.
4. Ethanol (absolute).
5. Isopropyl alcohol.

2.2.4. Purification of DNA From Cell Subsets Isolated by Flow-Sorting: Variant A, Qiagen Column Extraction

2.2.4.1. Hardware

1. Benchtop centrifuge.
2. Heat block or water bath.
3. Vortex mixer.

2.2.4.2. Reagents

1. Qiagen DNA Blood Mini Kit (Qiagen, Hilden, Germany), includes Protease, AL buffer, washing buffers AW1, AW2, and elution buffer AE.
2. Ethanol (absolute).

2.2.5. Purification of DNA From Cell Subsets Isolated by Flow-Sorting: Variant B, Proteinase K Lysis

2.2.5.1.Hardware

1. Eppendorf centrifuge.
2. Heat block or water bath.
3. Vortex mixer.

2.2.5.2. Reagents

1. 1 *M* Tris-HCl buffer, pH 8.0, molecular biology-grade, premade (Sigma).
2. 10 m*M* Tris-HCl pH 8.0 made from the above buffer; dilute the Tris with dH_2O only.
3. Proteinase K, molecular biology-grade (nuclease-free) (Boehringer); dilute to 100 µg/mL in 10 m*M* Tris-HCl, pH 8.0.

2.3. STR-PCR and Capillary Electrophoresis With Fluorescence-Assisted Detection (see **Note 1**)

2.3.1. Hardware

1. PCR cycler (PE2400, PE9600; Applied Biosystems, Foster City, CA).
2. Benchtop centrifuge.
3. Vortex mixer.
4. Capillary electrophoresis instrument equipped for analysis of fluorescence signals (ABI310/ABI 3100-Avant Genetic Analyzer, Applied Biosystems).

2.3.2. Reagents and Solutions

1. Genomic DNA.
2. 10X Buffer (including: Tris-HCl, KCl, $(NH_4)_2SO_4$, 15m*M* $MgCl_2$, pH 8,7) (Qiagen).
3. 25 m*M* $MgCl_2$ (Qiagen).
4. dNTPs (Invitrogen, Carlsbad, CA).
5. STR locus-specific primers (forward or reverse primer labeled with a fluorescence dye (FAM, HEX, or NED; Applied Biosystems) (*see* **Table 1**).
6. HotStar Taq Polymerase (Qiagen).
7. Deionized formamide (Applied Biosystems).
8. GeneScan-500 ROX Size Standard (Applied Biosystems).

2.3.3. Consumables for Capillary Electrophoresis on the ABI310/ABI 3100-Avant Genetic Analyzer

1. 10X GA Buffer (including EDTA) (Applied Biosystems).
2. 310 Genetic Analyzer Performance Optimized Polymer 4 (POP-4) or 3100 POP-4 Performance Optimized Polymer (Applied Biosystems).
3. 310 Genetic Analyzer Capillaries 47 cm × 50 μm or 3100-Avant Capillary Array 36 cm (Applied Biosystems).

3. Methods

3.1. Isolation of Cell Subsets by Flow-Sorting

1. Starting material is PB, usually anticoagulated with EDTA, or BM, usually anticoagulated with Heparin.
2. Determine WBC count using any available equipment and determine the PB/BM volume required for the forthcoming steps (*see* **Note 2**).
3. Transfer appropriate volume to a 50-mL tube and add 10 volumes of RBC lysis buffer.
4. Incubate in a refrigerator at +4°C until lysis is complete (5–10 min).
5. Centrifuge for 10 min at 400*g* in a refrigerated centrifuge at 4°C.
6. Remove supernatant and resuspend the pellet in 10 mL PBS.
7. Centrifuge for 10 min at 400*g* in a refrigerated centrifuge at 4°C.
8. Remove supernatant and resuspend the pellet of nucleated cells (NCs) in 1–2 mL PBS.
9. Determine the number of NC/μL using any available equipment.
10. Transfer 1X 10E6 NC for each MAb labeling reaction in a volume not exceeding 500 μL to a 15-mL tube for subsequent staining.
11. Add 60 μL MOPC-21 solution per 10^6 cells.
12. Vortex briefly.
13. Incubate the tubes in a refrigerator at 4°C for 20 min (*see* **Note 3**).
14. Add the respective MAb cocktails in a volume of 60 μL:
    a. Cocktail 1 (for flow-sorting of CD4, CD8, NK-cells, and normocytes): Syto41/ 8FITC/ 56PE/ 3ECD/ 45PerCP/ 4PE-Cy7/ 71APC/ 14APC-Cy7.

Table 1
Selected Microsatellite Loci

Locus	Marker	CHLC accession	Chromosome location	Type	Het	Alleles	PCR (in bp)	Primer sequences 5′>3′
D3S3045	GATA84B12	32627	3	TetNR	0.82	7	176–208	F: ACCAAATGAGACAGTGGCAT R: ATGAGGACGGTTGACATCTG
D4S2366	GATA22G05	31823	4	TetNR	0.79	7	120–144	F: TCCTGACATTCCTAGGGTGA R: AAAACAAATATGGCTCTATCTATCG
D12S1064	GATA63D12	40934	12	TetNR	0.82	8	173–201	F: ACTACTCCAAGGTTCCAGCC R: AATATTGACTTTCTCTTGCTACCC
D16S539	GATA11C06	715	16	TetNR	0.76	12	148–172	F: GATCCCAAGCTCTTCCTCTT R: ACGTTTGTGTGTGCATCTGT
D17S1290	GATA49C09	40873	17	TetNR	0.84	9	170–210	F: GCCAACAGAGCAAGACTGTC R: CGAAACAGTTAAATGGCCAA

The most informative set of short tandem repeat (STR) loci from the author's laboratory, which has provided an adequate marker in the majority of all donor/recipient constellations tested. A series of successful marker sets from different European laboratories has been recently published ***(6–11)***.

F, forward; R, reverse; Chrom, chromosomal; TetNR, tetranucleotide repeat marker; Het, heterozygosity. For the selection of appropriate markers for initial genotyping and subsequent analysis of chimerism, *see also* **Notes 19–23**.

   b. Cocktail 2 (for flow-sorting of granulocytes, monocytes, B lymphocytes and CD34 cells): Syto41/ 15FITC/33 PE/ 19ECD/ 45PerCP/ 34 PE-Cy7/ 71APC/ 14 APC-Cy7.
15. Vortex briefly.
16. Incubate in a refrigerator at 4°C for 30 min.
17. Remove excess MAb by adding approx 5 mL of RPMI culture medium containing 2% FCS and PS 1:100 vol.
18. Vortex briefly.
19. Centrifuge for 10 min at 400*g* in a refrigerated centrifuge at 4°C.
20. Remove supernatant and resuspend the pellet in 200 μL RPMI/2%FCS/PS.
21. Filter cell suspension through 40 μm mesh into a fluorescence-activated cell sorting (FACS) tube.
22. Isolate 4000 cells per population, whenever possible.
23. Collect individual cell populations in Eppendorf tubes (*see* **Note 4**).

3.2. Purification of DNA

3.2.1. DNA Extraction From PB or BM (see **Note 5**)

1. Use Qiagen DNA Blood Mini Kit according to the manufacturer's recommendations.
2. Quantify DNA yield by spectrophotometry.
3. Use 10 ng of DNA as template in individual PCR reactions.

3.2.2. DNA Extraction From Nails (see **Note 6**)

3.2.2.1. Variant A: In-House Method

1. Collect nail clippings from 1 to 10 fingers (or toes).
2. Place the nail sample is in a 1.5-mL Eppendorf tube.
3. Add washing solution.
4. Vortex briefly.
5. Pulse spin (Eppendorf centrifuge, maximum speed).
6. Aspirate washing solution.
7. Add approx 1 mL dH_2O.
8. Vortex briefly.
9. Pulse spin.
10. Aspirate dH_2O.
11. Repeat **steps 7–10** once.
12. Add extraction buffer.
13. Incubate sample at 56°C overnight (or over the weekend, if convenient).
14. Upon complete (or partial) digestion of the nails, extract the DNA using standard phenol/chloroform extraction and ethanol precipitation.

3.2.2.2. Variant B: Tissue and Hair Extraction Kit with DNA IQ System

The kits are applied according to the manufacturer's recommendations.

3.2.3. DNA Extraction From Flow-Sorted Cell Subsets (see **Note 7**)

3.2.3.1. Variant A: Qiagen Column Extraction

1. Resuspend flow-sorted cells with 200 µL PBS.
2. Add 20 µL protease.
3. Add 200 µl AL buffer.
4. Vortex for 15 s.
5. Incubate at 70°C for 10 min.
6. Pulse spin.
7. Add 210 µL ethanol (absolute).
8. Vortex briefly.
9. Pulse spin.
10. Apply the solution to the spin column.
11. Centrifuge and wash sample according to the standard Qiagen protocol.
12. Collect DNA with 100 µL of elution buffer (*see* **Note 8**).

3.2.3.2. Variant B: Proteinase K Lysis

1. Collect flow-sorted cells in 50 µL Tris-buffer.
2. Add 10 µL of 100 µg/mL proteinase K (PK).
3. Incubate for 1 h minimum at 56° C.
4. Vortex and spin down liquid briefly.
5. Incubate for 10 min at 95°C to inactivate the PK.
6. Vortex and spin down in Eppendorf centrifuge at full speed for 30 s to separate the DNA supernatant from the cellular debris.
7. Use 10–30 µL of the supernatant as template for the PCR reaction (avoid aspiration of the cellular debris in the pellet to prevent inhibition of the PCR reaction).

3.3. STR-PCR and Capillary Electrophoresis With Fluorescence-Assisted Detection for Chimerism Analysis (see **Note 9**)

1. Set up the PCR reactions in a total volume of 50 µL as follows:

dH20	5.5 µL
10X Buffer	5.0 µL
$MgCl_2$ (25 m*M*)	1.0 µL
dNTPs (from 2.5 m*M* stocks)	4.0 µL
Primers (from 2.5 pmol/µL stocks)	4.0 µL
Qiagen Taq Polymerase (from 5 units/µL stock)	0.5 µL
DNA (*see* **Note 10**)	30.0 µL
	50.0 µL

2. Perform PCR amplification under the following cycling conditions:

Initial denaturation at	95°C	15 min
32 cycles including denaturation at	94°C	60 s
annealing at	54°C	45 s
extension at	72°C	90 s
Final extension step at	72°C	7 min

Eliminate split peaks by adding an additional extension step at	60°C	45 min	(*see* **Note 11**)
Thereafter keep at until sample analysis by capillary electrophoresis (this temperature should prevent the dissociation of adenosine	18°C		(*see* **Note 11**)

3. Prepare PCR products for loading onto the capillary electrophoresis apparatus by setting up the following mixture:

13.7 μL	HiDi formamide
0.3 μL	ROX
1.0 LPCR	product

4. Denature the PCR products by incubating the tubes at 95°C for 2 min.
5. Let the samples cool down to room temperature prior to capillary electrophoresis.
6. Load samples to ABI 310/3100-Avant Genetic Analyzer (*see* **Note 12**).
7. Adjust the electrophoresis time ("run time") according to the length of the analyzed PCR products (*see* **Note 13**)
8. Apply the GeneScan software (supplied with ABI310/3100 apparatus) to analyze the PCR products.
9. Use the height (or area) of individual peaks to calculate donor/recipient chimerism by employing the formulas indicated below (*see* **Note 14**).

Commonly encountered problems and recommended measures are outlined in **Notes 15–18**.

Microsatellites for initial genotyping and selection of markers for the follow-up of chimerism are described in **Table 1** and **Notes 19–23**.

4. Notes

1. The main advantages of automated fluorescence-based detection of microsatellite markers over the use of conventional polyacrylamide gel electrophoresis (PAGE) include greater precision and easier performance of quantitative analysis, reduced manual handling of PCR products, and higher sensitivity.
2. We use a total of 4×10^6 NCs per patient sample for cell sorting. The total blood volume required is therefore based on the WBC count.
3. The incubation at 4°C minimizes nonspecific staining by blocking nonspecific antigen-binding regions.
4. If the subsequent DNA extraction is performed by PK lysis, the cells should be collected in a small amount of buffer, e.g., 50 μL Tris, as indicated in order to avoid the need for a DNA-concentrating step.
5. Capillary electrophoresis-based product analysis seems to be sensitive to variations in DNA template quality. It is necessary therefore to use isolation protocols yielding high quality DNA in order to obtain reproducible results and satisfactory sensitivity. DNA isolation kits (e.g., from Qiagen Inc) have proven to be adequate for this application.

6. Initial genotyping for the detection of informative STR loci is usually performed using PB from the recipient and PB or BM from the donor. In some instances, no cell material is available from the allograft recipient before transplantation. To identify an appropriate genetic marker for the monitoring of chimerism, assessment of the patient's genotype is necessary, but PB can no longer be used due to the presence of donor cells. We have tested cell material from different sources to identify patient-specific genetic fingerprints. Epithelial cells derived from the oral mucosa or from the skin may contain donor leukocytes. In fact, buccal swabs were demonstrated to contain granulocytes of donor origin already during the first days after SCT, before they were detected in PB (**ref. *16*** and own unpublished observations). The same applies to cells isolated from urinary sediment. Hair provide a reliable source of endogenous DNA, but may not be available in patients after several courses of chemotherapy. Nail clippings were found to be an ideal source of patient DNA in this setting. They are readily available in all patients, and sufficient quantities of good quality DNA adequate for PCR genotyping can be obtained from about 5 mg nail material. Usually, clippings of finger nails from both hands yield 20–50 mg of material, providing 10–25 μg of DNA by using the procedure indicated.
7. The amount of cells within individual fractions isolated by flow-sorting ranges mostly between 2000 and 10,000, but may occasionally be as low as a few hundred. A technique permitting efficient DNA extraction from small cell numbers is therefore required. We try to obtain 4000 cells per cell fraction in order to have sufficient amounts of DNA for PCR analysis. The use of a modified protocol for the Qiagen DNA Blood Mini Kit provides very pure DNA, but the yields are relatively low, albeit sufficient for subsequent PCR. The DNA extraction based on cell lysis and PK treatment is simpler, faster, and cheaper. The DNA yields are virtually quantitative, as determined by real-time PCR analysis of control genes, but the quality (purity) of DNA may not be adequate in all instances (i.e., may not work equally well with all primer combinations).
8. The volume of elution buffer should be kept low in order to provide DNA concentrations adequate for subsequent PCR analysis. The eluate can be re-applied to the spin column in attempts to in increase the DNA yield. Alternatively, if the yields do not provide sufficient amounts of DNA for PCR analysis, addition of a carrier to the cell lysate prior to application to the spin column may be warranted.
9. The employment of capillary electrophoresis instruments requires less hands-on time than conventional gel electrophoresis because of the automated loading of samples, electrophoresis, and measurement of fluorescence signals. However, the capacity of devices with a single capillary and an average electrophoresis time of 20–30 min per sample may be a limiting factor for sample throughput. The efficiency can be improved by loading PCR products of different chimerism assays onto the capillary and analyzing all fragments in the same run. This can be done when primers for amplification of various microsatellite loci are labeled with different dyes, thus permitting easy identification of the products after electrophoresis ***(17)***. These considerations are certainly of relevance at major diagnostic centers, where high sample throughput is required. Alternatively,

instruments with multiple capillaries can be used. Currently, the overall cost of capillary electrophoresis with fluorescence-based detection of PCR products is considerably higher than conventional analysis using PAGE.

10. The amount of template DNA in the PCR reactions can play a role in the achievable sensitivity of the assay. Most investigators use 50-100 ng of template (corresponding to about $7.5 \times 10^3 - 15 \times 10^3$ diploid human cells), a quantity readily available when analyzing chimerism within the entire WBC population. In such instances, it is feasible to reproducibly detect residual recipient populations in the range of 1% (i.e., about 100 cells). This level of sensitivity may not be readily achievable, however, when specific WBC fractions isolated by flow-sorting or by magnetic bead separation are investigated. In these instances, only small cell numbers are available for DNA isolation yielding no more than 1–30 ng of DNA (corresponding to about $1.5 \times 10^2 - 4.5 \times 10^3$ cells). Although 1% sensitivity (equivalent to detecting about 45 cells or less) can be reached even in this experimental setting, it is more common to achieve sensitivities around 3–5%. Despite the slightly decreased sensitivity of assays using low cell numbers as starting material, the overall sensitivity in detecting minor autologous cell fractions within specifically enriched leukocyte populations is generally one to two logs higher than chimerism analysis in whole WBC preparations.
11. Split peaks: several DNA polymerases can catalyze the addition of a single nucleotide (usually adenosine) to the 3′ends of double-stranded PCR amplicons. This nontemplate addition leads to the generation of PCR products that are one base pair longer than the actual target sequence. In order to avoid the occurrence of split peaks that are one base apart due to inefficient nucleotide addition, a terminal cycling step at 60°C for 45 min is included in the PCR profile. This step provides the polymerase with extra time to complete nucleotide addition to all double-stranded PCR products, thus usually preventing formation of split peaks.
12. The injection parameters need to be adjusted according to the yield of the PCR reaction and generally range between 5 and 15 s injection time and 1 and 6 kV. Usually, samples are run using different parameters in order to achieve optimal sensitivity. The peak height of the dominant alleles should be around 5000 rfu to permit detection of subdominant alleles at a sensitivity of around 1%, because the lower limit of detection (i.e., signal above noise) is around 50 rfu. Quantitative analysis is possible only if the detected peaks are not off-scale (indicated by the apparatus). In some instances, very high peaks (saturated signals) are not recognized and are indicated by the software as being off-scale, and appear as double-peaks (tip of the peak is bent down).
13. A run time of 20 min is sufficient for PCR products up to approx 400 bp in size; 30 min are required for longer products.
14. Quantification of the degree of mixed chimerism is often carried out relative to a patient-specific standard curve established from serial dilutions of pretransplant recipient in donor DNA. For each patient, standard curves are produced for one or more informative microsatellite markers. For quantitative analysis of donor and recipient alleles, both peak height and peak area can in principle be used. The formula used for calculating the degree of (most commonly recipient) chimerism

is based on the quotient between recipient and donor allele peak heights or areas. The mode of quantitative analysis may differ between individual centers ***(18)***. Some investigators select only one unique allele from each recipient and donor for the calculation Like many others, we include all unique recipient and donor alleles in the calculation, while shared alleles are excluded ***(19)***..The general formulas used is:

Percentage of recipient cells = 100×(A + B):(A + B) + (C + D)

The letters A and B represent the peak heights or areas of recipient alleles, the letters C and D those of donor alleles.

15. The so-called stutter peaks resulting from polymerase slippage during the amplification of microsatellite loci typically migrate at a distance of one repeat unit in front of the parent allele and may interfere with specific allelic peaks. This problem must be accounted for by judicious marker selection. As a general rule, informative donor and recipient alleles must be separated by at least two repeat units to prevent interference with stutter peaks. Occasionally, a second stutter allele, located at a distance of two repeat units from the main peak, may be present. If the second stutter is of relevant height, i.e. more than 1% of the main peak, another marker should be selected for chimerism analysis. The problem of second stutter peaks can be circumvented by selecting a marker, which provides donor and recipient peaks separated by more than two repeat units. In general, tetra- and pentanucleotide repeat markers (i.e., microsatellite loci displaying a repeat motif of four and five nucleotides in length, respectively), yield less prominent stutter peaks and are therefore preferred over di- and trinucleiotide repeat markers (i.e., microsatellite loci displaying a repeat motif of two or three nucleotides in length).
16. In certain instances, so-called "bleed-through" signals ***(20)*** may be observed, which may affect the interpretation of results. When using very high injection parameters, the ROX standard (red) may result in signals visible in analyses of PCR products labeled by HEX (green). The same phenomenon may occur in multiplex PCR reactions combining primers labeled by different dyes or in the presence of extremely high signals (*see* **Note 12**), where false (bleed-through) signals may be observed in different color windows.
17. The occurrence of dye-associated non-specific peaks ***(21)*** is a problem peculiar to the fluorescence-based technology discussed. Apparently, the fluorescent dyes FAM, HEX, and NED may give rise to formation of multiple template-independent peaks at positions characteristic for each dye. The peaks can be relatively high and migrate in a range similar to that of many microsatellite markers. The signals may therefore interfere with specific microsatellite peaks and thus compromise the analysis of chimerism. There is currently no clear explanation for this phenomenon. If this problem is observed, a feasible approach to its elimination is having primers for each microsatellite locus labeled with different fluorescent dyes and selecting a primer/dye combination that does not show any interference between the specific alleles of the marker used and the dye-associated peak positions.

18. In addition to the occurrence of nonspecific peaks, as outlined in **Notes 11** and **15–17**, other problems may lead to the appearance of extra signals. Conversely, the problem of faint or absent peaks may also occur. Some of the common causes and possible measures are listed in **Table 2**.
19. Only a limited number of highly polymorphic microsatellite loci need to be tested to provide an informative marker in virtually all donor/recipient constellations. In matched sibling and haploidentical transplants, more markers are generally required in order to differentiate between donor and recipient T-cells as compared to the unrelated setting. The screening panels used at most centers include 6–15 microsatellite markers. Genotyping with a panel of 3–7 markers is often sufficient to reveal allelic differences between recipient- and donor-derived cells suitable for the analysis of chimerism **(Table 1)**.
20. The criteria for the selection of informative markers are not uniformly defined ***(22)***. Ideally, at least one unique donor and recipient allele should be present to render a marker eligible for chimerism testing. For the monitoring of residual recipient hemopoiesis only, the presence of a unique recipient allele distinguishable from the donor allele(s) could be regarded as the minimum requirement.
21. As a result of preferential amplification of short fragments during PCR cycling, minor recipient cell populations may be detected with greater sensitivity if the informative recipient alleles are shorter in length than any donor allele. Although microsatellite markers usually show relatively small differences in length between alleles, amplification efficiencies may nevertheless differ significantly and therefore have an impact on the sensitivity of the assays.
22. Investigation of a posttransplant DNA sample with multiple microsatellite markers may improve the reproducibility and accuracy of quantitative chimerism analysis. The accuracy of quantitative chimerism assays can be increased by testing each sample with more than one marker and calculating mean values ***(19)***. Some investigators therefore perform clinical testing of chimerism with commercial multiplex kits facilitating co-amplification of several microsatellite markers in a single PCR reaction ***(17)***. However, the advantages indicated previously are counterbalanced by significantly higher cost of consumables and lower sensitivity resulting from the high number of different fragments co-amplified. Most diagnostic centers therefore rely on the use of singleplex PCR reactions for chimerism analysis. Multiplex PCR assays are sometimes used for initial recipient/donor genotyping to select one or more informative markers for the monitoring of chimerism. If posttransplant DNA samples are tested by more than one microsatellite marker, amplification is usually performed in separate PCR reactions.
23. The reported average sensitivity in detecting minor (in most instances recipient-derived) cell populations using different microsatellite markers ranges from 1 to 5%. The sensitivity may to a large extent depend on the size (i.e., amplicon length) of the informative recipient allele(s), the allelic constellation and the number of alleles co-amplified. In practice, however, some markers from the panels used at individual centers tend to provide higher sensitivity than others and are therefore used preferentially.

Table 2
Commonly Encountered Problems in STR-PCR and Recommended Measures

	Possible cause	Recommended measure
Extra signals:	• Contamination with extraneous DNA	• Use appropriate precautions and controls (aerosol-resistant pipet tips)
	• DNA input too high	• Decrease amount of template DNA or reduce the number of PCR cycles
	• Sample not completely denatured	• Heat sample to 93°C for 3 min before capillary electrophoresis
Faint or no signals:	• Impure DNA: Presence of inhibitors	• Test different extraction methods combined with filtration of DNA through a spin column
	• Insufficient template DNA	• Increase amount of DNA template
	• Primer concentration too low	• Increase primer concentration
	• Incorrect PCR program	• Check the PCR program
	• Wrong $MgCl_2$ concentration	• Test different concentrations of $MgCl_2$

Acknowledgments

Supported by the "Österreichische Kinderkrebshilfe." The following colleagues from the CCRI, Vienna, Austria have contributed to the contents of this chapter: Dieter Printz and Gerhard Fritsch (flow-sorting), Helga Daxberger, and Sandra Preuner (STR-PCR and capillary electrophoresis).

References

1. Dubovsky, J., Daxberger, H., Fritsch, G., et al. (1999) Kinetics of chimerism during the early post-transplant period in pediatric patients with malignant and non-malignant hematologic disorders: implications for timely detection of engraftment, graft failure and rejection. *Leukemia* **13,** 2060–2069.
2. Pérez-Simón, J. A., Caballero, D., Lopez-Pérez, R., et al. (2002) Chimerism and minimal residual disease monitoring after reduced intensity conditioning (RIC) allogeneic transplantation. *Leukemia* **16,** 1423–1431.
3. Matthes-Martin, S., Lion, T., Haas, O. A., et al. (2003) Lineage-specific chimaerism after stem cell transplantation in children following reduced intensity conditioning: potential predictive value of NK-cell chimaerism for late graft rejection. *Leukemia* **17,** 1934–1942.
4. Lion, T., Daxberger, H., Dubovsky, J., et al. (2001) Analysis of chimersim within specific leukocyte subsets for detection of residual or recurrent leukemia in pediatric patients after allogeneic stem cell transplantation. *Leukemia* **15,** 307–310.
5. Bader, P., Kreyenberg, H., Hoelle, W., et al. (2004) Increasing mixed chimerism is an important prognostic factor for unfavorable outcome in children with acute lymphoblastic leukemia after allogeneic stem-cell transplantation: Possible role for pre-emptive immunotherapy? *J. Clin. Oncol.* **22,** 1696–1705.
6. Chalandon, Y., Vischer, S., Helg, C., Chapuis, B., and Roosnek, E. (2003) Quantitative analysis of chimerism after allogeneic stem cell transplantation by PCR amplification of microsatellite markers and capillary electrophoresis with fluorescence detection: the Geneva experience. *Leukemia* **17,** 228–231.
7. Schraml, E., Daxberger, H., Watzinger, F., and Lion, T. (2003) Quantitative analysis of chimerism after allogeneic stem cell transplantation by PCR amplification of microsatellite markers and capillary electrophoresis with fluorescence detection: the Vienna experience. *Leukemia* **17,** 224–227.
8. Kreyenberg, H., Holle, W., Mohrle, S., Niethammer, D., and Bader, P. (2003) Quantitative analysis of chimerism after allogeneic stem cell transplantation by PCR amplification of microsatellite markers and capillary electrophoresis with fluorescence detection: the Tuebingen experience. *Leukemia* **17,** 237–240.
9. Acquaviva, C., Duval, M., Mirebeau, D., Bertin, R., and Cave, H. (2003) Quantitative analysis of chimerism after allogeneic stem cell transplantation by PCR amplification of microsatellite markers and capillary electrophoresis with fluorescence detection: the Paris-Robert Debre experience. *Leukemia* **17,** 241–246.
10. Hancock, J. P., Goulden, N. J., Oakhill, A., and Steward, C. G. (2003) Quantitative analysis of chimerism after allogeneic bone marrow transplantation using

immunomagnetic selection and fluorescent microsatellite PCR. *Leukemia* **17,** 247–251.

11. Koehl, U., Beck, O., Esser, R., et al. (2003) Quantitative analysis of chimerism after allogeneic stem cell transplantation by PCR amplification of microsatellite markers and capillary electrophoresis with fluorescence detection: the Frankfurt experience. *Leukemia* **17,** 232–236.
12. Maas, F., Schaap, N., Kolen, S., et al. (2003) Quantification of donor and recipient hemopoietic cells by real-time PCR of single nucleotide polymorphisms. *Leukemia.* **17,** 621–629.
13. Fredriksson, M., Barbany, G., Liljedahl, U., Hermanson, M., Kataja, M., Syvänen, A. C. (2004) Assessing hematompoietic chimerism after allogeneic stem cell transplantation by multiplexed SNP genotyping using microarrays and quantitative analysis of SNP alleles. *Leukemia* **18,** 255–266.
14. Jimenez-Velasco, A., Barrios, M., Roman-Gomez, J., et al. (2005). Reliable quantification of hematopoietic chimerism after allogeneic transplantation for acute leukemia using amplification by real-time PCR of null alleles and insertion/deletion polymorphisms. *Leukemia* **19,** 336–343.
15. Najfeld, V., Burnett, W., Vlachos, A., Scigliano, E., Isola, L., and Fruchtman, S. (1997) Interphase FISH analysis of sex-mismatched BMT utilizing dual color XY probes. *Bone Marrow Transplant.* **19,** 829–834.
16. Thiede, C., Prange-Krex, G., Freiberg-Richter, J., Bornhauser, M., and Ehninger, G. (2000) Buccal swabs but not mouthwash samples can be used to obtain pretransplant DNA fingerprints from recipients of allogeneic bone marrow transplants. *Bone Marrow Transplant.* **25,** 575–577.
17. Thiede, C., Bornhäuser, U., Brendel, C., et al. (2001) Sequential monitoring of chimerism and detection of minimal residual disease after allogenic blood stem transplantation (BSCT) using multiplex PCR amplification of short tandem repeat-markers. *Leukemia* **15,** 293–302.
18. Lion, T. (2003) Summary: Reports on quantitative analysis of chimaerism after allogeneic stem cell transplantation by PCR amplification of microsatellite markers and capillary electrophoresis with fluorescence detection. *Leukemia* **17,** 252–254.
19. Thiede, C. and Lion, T. (2001) Quantitative analysis of chimerism after allogeneic stem cell transplantation using multiplex PCR amplification of short tandem repeat markers and fluorescence detection. *Leukemia* **15,** 303–306.
20. Moretti, T. R., Baumstark, A. L., Defenbaugh, D. A., Keys, K. M., Smerick, J. B., and Budowle, B. (2001) Validation of short tandem repeats (STRs) for forensic usage: performance testing of fluorescent multiplex STR systems and analysis of authentic and simulated forensic samples. *J. Forensic Sci.* **46,** 647–660.
21. Schraml, E. and Lion, T. (2003) Interference of dye-associated fluorescence signals with quantitative analysis of chimerism by capillary electrophoresis. *Leukemia* **17,** 221–223.
22. Thiede, C., Borhäuser, M., and Ehninger, G. (2004) Evaluation of STR informativity for chimerism testing-comparative analysis of 27 STR systems in 203 matched related donor recipient pairs. *Leukemia* **18,** 248–254.

15

Application of Proteomics to Posttransplantational Follow-Up

Eva M. Weissinger and Harald Mischak

Summary

Proteomic screening of complex biological samples becomes of increasing importance in clinical research and diagnosis. It is expected that the meager number of approx 35,000 human genes gives rise to more than 1,000,000 functional entities at the protein level. Thus, the proteome provides a much richer source of information than the genome for describing the state of health or disease of the human organism. Especially, the composition body fluids comprise a rich source of information on possible changes in the status of health or disease of particular organs and in consequence of the whole organism. Here we describe the application of capillary electrophoresis (CE) coupled on-line to an electrospray-ionization (ESI)-time-of-flight (TOF)-mass spectrometer) to the analysis of human urine for the identification of biomarkers for complications after allogeneic hematopoietic stem cell transplantation.

Key Words: Proteomics; hematopoietic stem cell transplantation; capillary electrophoresis; mass spectrometry; polypeptide; clinical diagnosis.

1. Introduction

Proteome analysis is now emerging as key technology for deciphering biological processes and the discovery of biomarkers for diseases from tissues and body fluids in biosciences and increasingly in clinical research. The complexity and wide dynamic range of protein expression poses an enormous challenge to both separation technologies as well as consequent detection such as mass spectrometry (MS). The combination of the high resolution separation properties of capillary electrophoresis (CE) with the sensitive and accurate mass identification of MS to date fulfills most of these requirements. Several approaches for CE-MS coupling have been published ***(1–3)***. Recently developed MS coupling techniques ***(4)*** lead to robust CE-MS applications with sensitivity in the

From: *Methods in Molecular Medicine, vol. 134: Bone Marrow and Stem Cell Transplantation*
Edited by: M. Beksac © Humana Press Inc., Totowa, NJ

high atto-mol range, comparable to the nano-liquid chromatography systems ***(5)***. Consequently, CE-MS offers a fast, accurate, and reproducible system for the analysis of clinical samples. More than 1000 individual proteins and peptides can be detected in a single CE-MS run within 45 to 60 min, thus fulfilling the requirements for analysis in a clinical setting ***(6)***. The clinical follow up of urine samples of patients after allogeneic hematpoietic stem cell transplantation (HSCT) *(7)* will be used to describe the method.

2. Materials

1. Urine samples collected cold and immediately frozen and stored at –20°C.
2. Buffer for sample dilution: 4 *M* urea, 20 m*M* NH_4OH containing 0.2% sodium dodecyl sulfate (SDS).
3. Amicon Ultra centrifugal filter device (30 kDa; Millipore, Bedford, MA).
4. C2-column (Pharmacia, Uppsala, Sweden).
5. Elution buffer: 50% acetonitrile (ACN) in high-performance liquid chromatography (HPLC)-grade H_2O containing 0.5% formic acid.
6. Lyophilization Christ Speed-Vac RVC 2-18/Alpha 1-2 (Christ, Osterode am Harz, Germany).
7. Capillary Electrophoresis P/ACE MDQ system (Beckman Coulter, Fullerton, CA).
8. Capillary: 90 cm, 50 μm I.D. fused silica capillary.
9. Rinse between runs: 0.1 *M* NaOH.
10. CE-Running buffer: 20% ACN, 0.25 *M* formic acid.
11. Sheath liquid 30% (v/v) iso-propanol (Sigma-Aldrich, Germany) and 0.4% (v/v) formic acid in HPLC-grade water (flow rate: 2 μL/min).
12. Electrospray ionization (ESI)-time-of-flight (TOF) sprayer-kit from (Agilent technologies, Palo Alto, CA).
13. Micro-TOF mass spectrometer (ESI-TOF; Bruker Daltonik, Bremen, Germany).
14. Software: Mosaiques Visu: for peak detection (Biomosaiques software, Hannover, Germany)
15. MosaCluster: SVM-based program for grouping according to multiple parameters. (Biomosaiques software,Hannover, Germany).

3. Methods

The methods described below outline (1) the sample preparation, (2) the CE-MS analysis, and (3) data processing and analysis to evaluate the abundant information obtained after a single run in the CE-MS for clinical application to obtain particular patterns typical for different diseases or complications.

3.1. Sample Preparation

Body fluids are complex mixtures of molecules with a wide range of polarity, hydrophobicity, and size. When analyzing complex biological samples, major concerns are loss of analytes (here: polypeptides) during the sample preparation as well as reproducibility of the data generated. Ideally, a crude, unprocessed sample should be analyzed. This would avoid all artefacts, losses, or biases arising from sample preparation. To date the presence of large molecules, such as albumin, immunoglobulin, and others hamper this direct approach, because these molecules bare little information in the clinical setting, but will interfere with the detection of smaller, less abundant proteins and peptides. Thus a manipulation of the samples is necessary, but can be limited to a few steps.

3.1.1. Collection and Storage of the Samples

Urine samples are obtained from patients at different time points before and after HSCT, starting prior to conditioning and consequently weekly for the time on the ward. The samples are typically taken as the second spot urine, voiding the first urine of the day. Proper handling of the samples is of outmost importance; the urine should be frozen immediately after collection and stored at –20°C until analysis (**Note 1**).

3.1.2. Sample Preparation

Shortly before analysis, 1-mL aliquots are thawed and diluted with 1 mL of 4 *M* Urea, 20 m*M* NH_4OH containing 0.2% SDS. This is followed by an ultrafiltration step using the Amicon Ultra centrifugal filter device (30 kDa; Millipore, Bedford, MA). Samples are spun at 3,000*g* until 1.5 mL of filtrate has passed through the filter, thus removing the high-molecular-weight proteins such as albumin, transferrin and others (**Note 2**). The filtrate is then desalted and applied onto a reversed phase C2-column to remove urea, salts, and other confounding materials. Polypeptides are eluted with 50% ACN in HPLC-grade H_2O with 0.5% formic acid. Next, samples are lyophilized over night and resuspended in 50 μL of HPLC-grade H_2O shortly before injection.

3.2. CE-MS Analysis of the Samples

3.2.1. Capillary Electrophoresis

The samples are transferred to appropriate vials and stored in the CE-auto sampler section at 5°C. For capillary electrophoresis a P/ACE MDQ (Beckman Coulter, Fullerton, CA) system is used equipped with a 90-cm, 50-μm inner diameter, bare-fused silica capillary. The use of coated capillaries appears not to be beneficial (**Note 3**). The capillary is first rinsed with running buffer (20% ACN, 0.5% formic acid, 79.5% HPLC-grade water) for 3 min prior to sample injection. The sample is injected for 99 s with 1–6 psi, resulting in the injection of 60–300 nL sample. Separation is performed with +30 kV at the injection side and the capillary temperature is set to 35°C for the entire length of the capillary

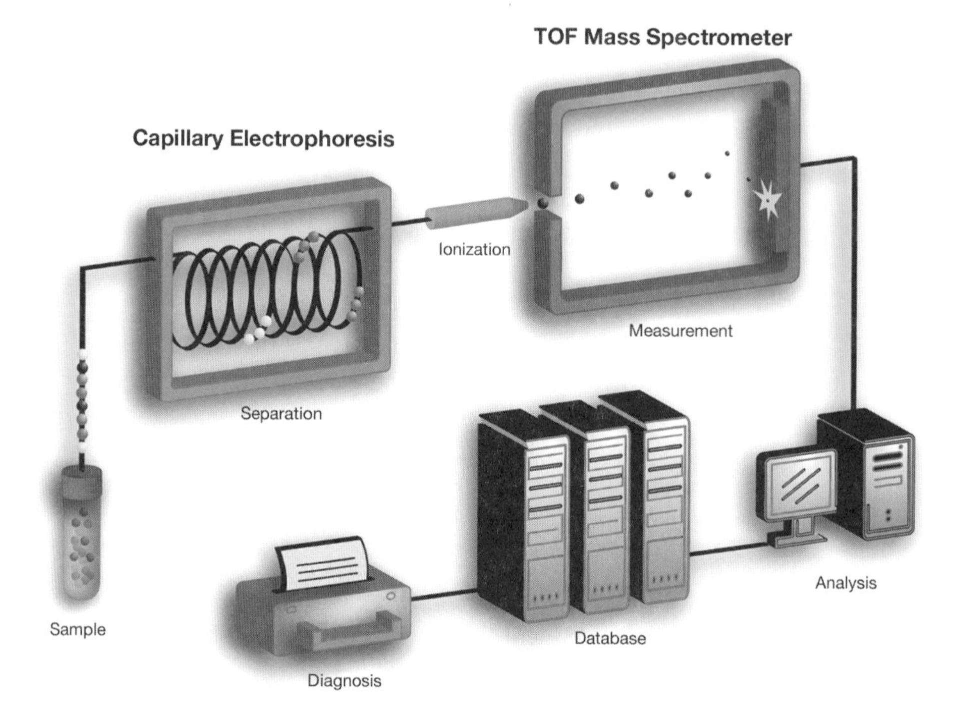

Fig. 1. On line coupling of capillary electrophoresis (CE)-mass spectrometry (MS) and setup for data processing. A schematic drawing of the on-line coupling of CE-MS beginning at the preparation of the sample preparation to the final data processing and identification of the pattern is shown here.

up to the ESI interface. After each run, the capillary is rinsed for 5 min with 0.1 *M* NaOH to remove protein build up that accumulates if uncoated capillaries are used **(Note 3)**. This is followed by a 5-min rinse with HPLC-grade H_2O and subsequently with running buffer. The CE-MS set up is depicted in **Fig. 1**.

3.2.2. CE-MS Interface and Analysis

The MS analysis is performed in positive electrospray mode with an ESI-TOF sprayer-kit from (Agilent technologies, Palo Alto, CA) using a Micro-TOF mass spectrometer (Bruker Daltonik, Bremen, Germany). The ESI sprayer is grounded and the ionspray interface potential is set between –3700 and –4100 V. The sheath liquid is applied coaxially, consisting of 30% (v/v) iso-propanol (Sigma-Aldrich) and 0.4% (v/v) formic acid in HPLC-grade water, at a flow rate of 1–2 μL/min.

These conditions result in a detection limit of about 1 fmol of different standard peptides *(6)*. MS-spectra are accumulated every 3 s, over a mass-to-charge range from 400 to 2500 or 3000 m/z for about 45-60 min.

3.3. Data Processing and Statistical Analysis

3.3.1. Data Processing: Peak Annotation

m/z values of MS peaks are deconvoluted into mass and combined if they represented identical molecules at different charge states using MosaiquesVisu *(8)*. MosaiquesVisu (accessible at www.proteomiques.com) employs a probabilistic clustering algorithm and uses both isotopic distribution as well as conjugated masses for charge-state determination of polypeptides. In a first electronic analysis, the software identifies all peaks within each single spectrum. This usually results in more than 100,000 peaks from a single sample. Because true analytes must appear in several, e.g., at least three, successive spectra, signals from individual peptides present in consecutive spectra are collected, evaluated with respect to charge state and combined in the next step. This reduces the list to 3000–7000 "CE/MS" peaks. These data are termed "peak list" of an individual sample. The peak list can be converted into a three-dimensional plot, m/z on the *y*-axis plotted against the migration time on the *x*-axis and the signal intensity color-coded **(Fig. 2)**. Signals must meet certain criteria, such as: I(1) the signal intensity must be greater then the threshold (usually S/N > 7); (2) single-charged peaks are discarded; and/or (3) the peak's width must be below the threshold of 2 min. Molecules that do not meet these criteria are not proteins and are removed from the list.

3.3.2. Data Processing: Peak Deconvolution, Protein Contour Plots

In the next step of the data processing, peaks representing identical molecules with different charge states are identified. These are deconvoluted into a single mass. The actual mass of the polypeptides plotted against the migration time result in the "protein-plot" **(Fig. 2)**. This theoretical CE-MS spectrum that now contains the information on migration time, mass, and signal intensity for each individual polypeptide generally consists of between 1000 and 2500 individual polypeptides per sample. To allow comparison and search for conformity and differences between the samples obtained at different time points after HSCT and from different individuals, CE-migration times have to be normalized using ca 200 polypeptides generally present in a urine sample that serve as internal standards *(9)*. The signal intensity is normalized to the total ion current (TIC) of the utilized signals. Polypeptides within different samples are considered identical if the mass deviation is less than 50 ppm and the CE migration-time deviation is less than 2 min.

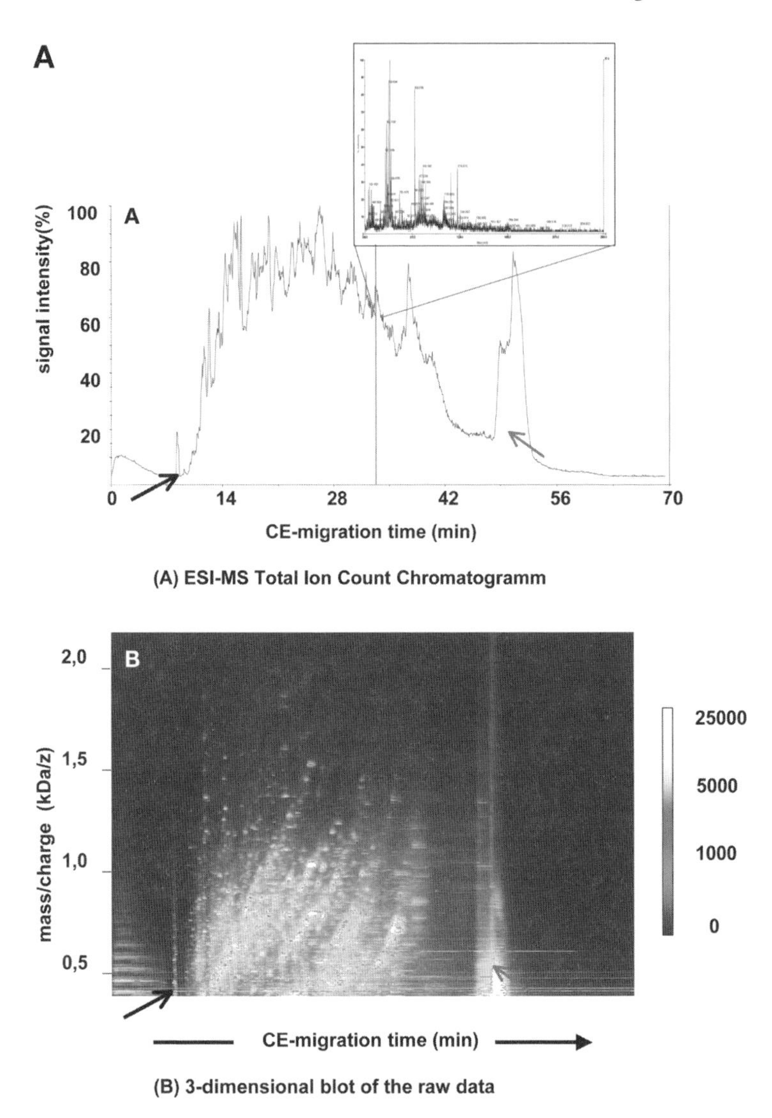

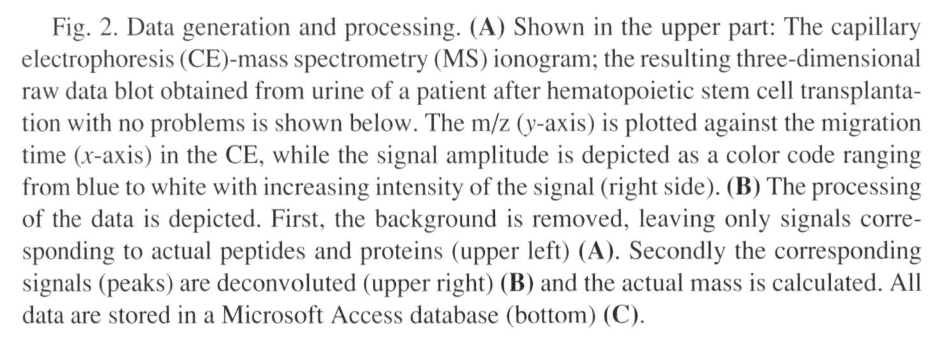

Fig. 2. Data generation and processing. **(A)** Shown in the upper part: The capillary electrophoresis (CE)-mass spectrometry (MS) ionogram; the resulting three-dimensional raw data blot obtained from urine of a patient after hematopoietic stem cell transplantation with no problems is shown below. The m/z (y-axis) is plotted against the migration time (x-axis) in the CE, while the signal amplitude is depicted as a color code ranging from blue to white with increasing intensity of the signal (right side). **(B)** The processing of the data is depicted. First, the background is removed, leaving only signals corresponding to actual peptides and proteins (upper left) **(A)**. Secondly the corresponding signals (peaks) are deconvoluted (upper right) **(B)** and the actual mass is calculated. All data are stored in a Microsoft Access database (bottom) **(C)**.

B

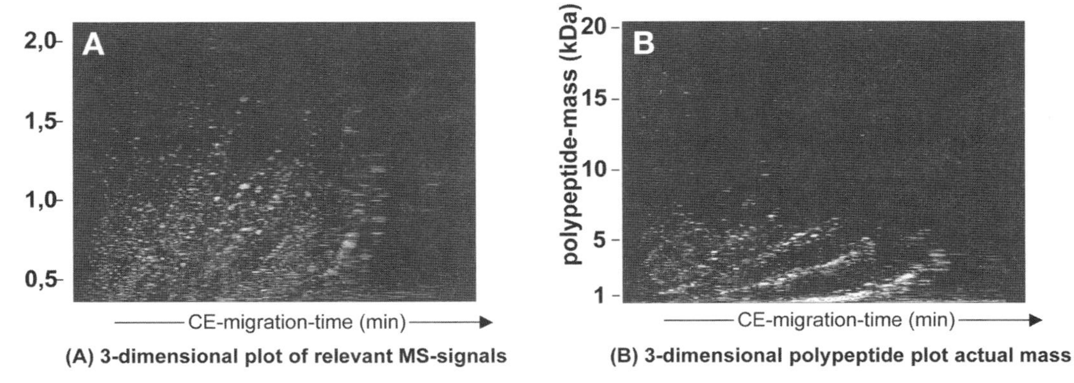

(A) 3-dimensional plot of relevant MS-signals

(B) 3-dimensional polypeptide plot actual mass

C

34.45±0.23	638.00	1911.01	3 (97 %)	3 (96 %)	532036	34.44_1911.0	inf
34.45±0.25	956.48	1910.97	2 (94 %)	2 (99 %)	291315	34.44_1911.0	inf
17.60±0.28	950.58	4747.92	5 (0 %)	5 (90 %)	88627	17.60_4748.2	inf
39.41±0.26	1153.77	3458.31	3 (62 %)	3 (68 %)	77940	39.39_3457.7	inf
38.24±0.18	726.80	1451.61	2 (98 %)	2 (98 %)	62442	38.25_1451.6	inf
33.85±0.79	543.31	542.31	1 (92 %)	1 (100 %)	40451	33.85_542.3	inf
17.62±0.21	792.34	4748.05	6 (0 %)	6 (92 %)	39940	17.60_4748.2	inf
34.39±0.36	791.43	1580.85	2 (94 %)	2 (79 %)	37025	34.16_1580.9	inf
35.73±0.85	1235.23	6171.17	3 (0 %)	5 (91 %)	35453	35.67_6171.0	inf
34.27±0.50	734.88	1467.77	2 (98 %)	2 (100 %)	33761	34.27_1467.8	inf
27.93±0.28	680.70	2039.11	3 (91 %)	3 (98 %)	32946	27.93_2039.1	inf
39.93±0.19	475.24	474.24	1 (100 %)	1 (100 %)	32547	39.93_474.2	inf
34.46±0.25	632.00	1893.01	3 (93 %)	3 (100 %)	28308	34.46_1893.0	inf
26.67±0.28	613.36	1224.72	2 (97 %)	2 (100 %)	27073	26.67_1224.7	inf
33.64±0.23	942.71	2825.12	3 (75 %)	3 (99 %)	26521	33.65_2825.1	inf
49.40±0.42	762.82	1523.65	2 (97 %)	2 (100 %)	25937	49.40_1523.6	inf
36.02±0.67	1238.33	6186.67	3 (0 %)	5 (87 %)	25850	36.15_6186.7	inf
33.71±0.40	656.40	655.40	1 (99 %)	1 (100 %)	23127	33.71_655.4	inf
39.82±0.18	449.24	448.24	1 (100 %)	1 (100 %)	22789	39.82_448.2	inf
33.45±0.45	883.50	882.50	1 (89 %)	1 (81 %)	21883	33.61_882.5	inf
39.69±0.21	878.45	1754.91	2 (91 %)	2 (84 %)	21085	39.73_1754.8*	inf
44.80±0.50	609.79	1217.57	2 (96 %)	2 (89 %)	21016	44.75_1217.5	inf
19.57±0.23	831.63	3322.51	4 (54 %)	4 (91 %)	19586	19.54_3322.6	inf
39.36±0.20	1148.46	3442.37	3 (60 %)	3 (85 %)	18737	39.37_3442.1	inf
28.09±0.40	683.61	2047.83	3 (86 %)	3 (81 %)	18160	28.21_2047.7	89
41.31±0.23	692.34	691.34	1 (99 %)	1 (100 %)	18129	41.31_691.3	inf
17.60±0.24	1188.15	4748.61	4 (0 %)	4 (86 %)	17985	17.60_4748.2	inf
22.58±1.03	851.64	3402.56	4 (52 %)	4 (71 %)	17858	22.38_3402.9	85

34.44_1911.0	27.21*	1910.97	178712*	100%	3	1-3	inf
17.60_4748.2	10.88*	4748.20	29964*	100%	5	3-7	inf
39.39_3457.7	32.01*	3457.66	14519*	98%	3	2-4	inf
38.25_1451.6	30.90*	1451.60	13732*	100%	2	1-2	inf
35.67_6171.0	28.40*	6170.96	10256*	100%	7	3-6	82
33.85_542.3	26.64*	542.31	9052*	92%	1	1	inf
27.93_2039.1	20.89*	2039.08	7566*	100%	2	2-3	inf
34.27_1467.8	27.04*	1467.77	7555*	98%	1	2	inf
39.93_474.2	32.53*	474.24	7283*	100%	1	1	inf
33.65_2825.1	26.44*	2825.11	6956*	100%	3	2-4	inf
34.16_1580.9	26.93*	1580.87	6892*	100%	2	2-3	inf
34.46_1893.0	27.22*	1893.01	6334*	93%	1	3	inf
36.15_6186.7	28.86*	6186.70	6168*	100%	4	3-6	73
26.67_1224.7	19.67*	1224.72	6058*	97%	1	2	inf
49.40_1523.6	41.71*	1523.65	5804*	97%	1	2	inf
33.61_882.5	26.40*	882.53	5389*	100%	2	1-2	95
29.99_3209.2	22.89*	3209.22	5246*	100%	3	3-5	69
33.71_655.4	26.50*	655.40	5175*	99%	1	1	inf
35.53_1046.5	28.26*	1046.52	5141*	100%	2	1-2	inf
39.82_448.2	32.42*	448.24	5100*	100%	1	1	inf
44.75_1217.5	37.20*	1217.53	4813*	100%	2	1-2	inf
19.54_3322.6	12.76*	3322.59	4789*	100%	3	3-5	inf
18.89_3840.6	12.13*	3840.63	4693*	100%	4	3-5	98
39.37_3442.1	31.98*	3442.06	4312*	100%	3	2-4	66
41.31_691.3	33.87*	691.34	4057*	99%	1	1	inf
39.73_1754.8	32.33*	1754.83	4008*	100%	3	1-2	inf
34.49_1932.9	27.25*	1932.95	3935*	100%	2	2-3	73
21.84_6212.9	14.98*	6212.90	3741*	100%	4	4-7	85

(C) Deconvolution of MS peaks (A) to one actual mass (B)

Intra- and interassay variability are ascertained by repeated analysis of one sample and by analysis of samples obtained at different time points from the same patient under comparable conditions, respectively.

3.3.3. Polypeptide Patterns for Prediction of Complications After Stem Cell Transplantation

Complications after stem cell transplantation can be predicted with high significance by screening patient's urine routinely with CE-MS. Urine from 50 patients after hematopoietic stem cell transplantation (HSCT; 45 after allogeneic and 5 after autologous HSCT) and 8 with sepsis were collected up to day +365 after HSCT. Screening led to the generation of polypeptide patterns yielding an early recognition of changes, if complications occurred during the observation period. 20 patients developed graft-vs-host disease (GvHD) after allo-HSCT. The polypeptide patterns yielded differentially excreted, statistically relevant polypeptides **(Fig. 3)** forming a pattern specific for recognition of GvHD. Comparison with patients with sepsis allowed to distinguish sepsis from GvHD with a specificity 97% and a sensitivity of 98% based on an inclusion of sepsis-specific polypeptides. Thus, the application of CE-MS to the screening of patients after HSCT can help to reduce transplantation related morbidity and mortality significantly.

4. Notes

1. The collection of the samples under proper conditions is of extreme importance for sensitive analysis such as MS. Whereas in Western blot or ELISA minor degradation of the proteins generally does not cause problems, CE-MS or actually any MS-based proteomic analysis requires storage of the samples at –20°C immediately after collection, in order to prevent degradation of proteins, which would result in additional fragments of proteins that are probably irrelevant to the underlying disease. Collecting the samples on ice is another measure to prevent degradation. Keeping the sample cold and immediate freezing is preferable over addition of protease inhibitors.
2. Initial experiments and also our previous data ***(10)*** revealed that higher molecular weight proteins in the urine sample generally appear not to contain significant information, but cause severe problems during the CE-MS runs, such as precipitation, clogging of the capillary, and overloading with respect to the smaller, more interesting molecules. Therefore, an optimized protocol to remove those using ultrafiltration has been developed. The most abundant higher molecular weight protein is albumin, a carrier protein that binds a substantial fraction of other proteins and peptides. In order to prevent this interaction, we use a chaotropic agent, urea, in combination with a detergent, SDS (dilution of the samples). Consistently good and reproducible results and recovery of {GT}80% of added standard polypeptides were observed in the presence of 2 *M* urea and 0.1% SDS ***(11)***.

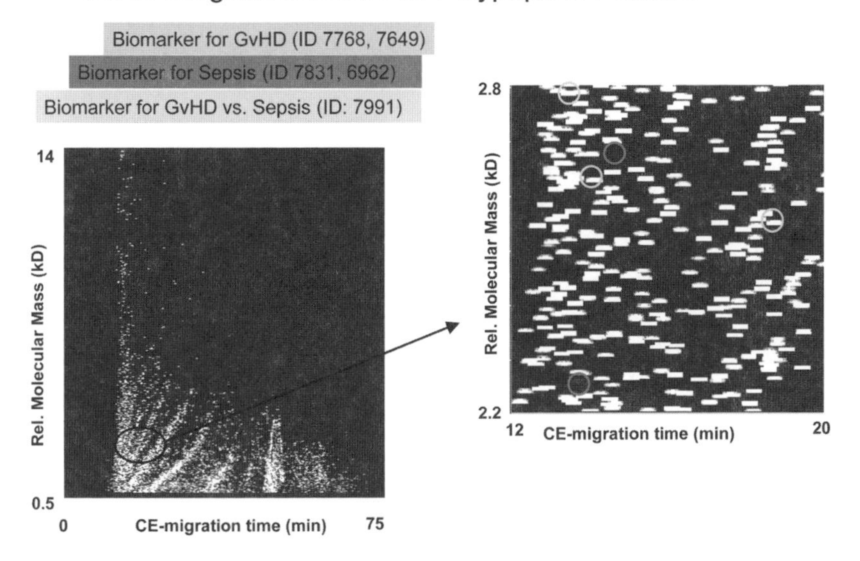

Fig. 3. Diagnosis of graft-vs-host disease (GvHD) in a biological sample. In a first set of analysis, the biomarker pattern specific for complications after hematopoietic stem cell transplantation (such as sepsis, GvHD, and others) are established. Next, all other samples from patients are searched for the presence of particular markers for any of these complications. The presence or absence of each marker is scored and stored in the database.

3. Under neutral or basic conditions the silanol groups of the capillary form a negatively charged surface which interacts with the positive charge of the proteins. This leads to peak broadening especially of highly charged molecules. To circumvent these problems, several different coatings and coating protocols have been described ***(12–14)***. Hence, we have examined several mostly silyl-based coatings ***(15)***. Unfortunately, most of the coating procedures described in the literature are quite tedious and time-consuming. Coated capillaries gave good results when using standard polypeptides. However, coatings that were examined in our laboratory were not satisfactory with "real" clinical samples such as urine, because these appear to deposit in part on the coating material, which leads to peak broadening and increased migration times. In a very recent manuscript Ullsten et al. ***(16)*** describe a new coating that seems to represent a major improvement. This polyamine coating (PolyE-323) was reported to show good stability even at high pH, involves no complicated chemical reactions, and can also be applied onto the capillary within minutes. It still remains to be seen whether this type of coating is compatible with clinical samples, but it appears to be a very promising approach.

Consequently, our experience using different types of coating so far were quite unsatisfactory and the optimal approach appeared to be the use of uncoated capillaries. Decreasing the pH of the background electrolyte (BGE) reduces the negative charge of the surface, thus reducing capillary-protein-interaction. Additionally, hydrophobic interactions can be reduced by adding organic solvent to the BGE. Therefore, best results are obtained at low pH in the presence of acetonitrile.

Last, it is important to point out that most coatings result in a positive charged capillary wall and cause an inverse electro-osmotic flow; consequently, the electrical field must be reversed.

References

1. Manabe, T. (1999) Capillary electrophoresis of proteins for proteomic studies. *Electrophoresis* **20,** 3116–3121.
2. Oda, R. P., Clark, R., Katzmann, J. A., and Landers, J. P. (1997) Capillary electrophoresis as a clinical tool for the analysis of protein in serum and other body fluids. *Electrophoresis* **18,** 1715–1723.
3. Jellum, E., Dollekamp, H., and Blessum, C. (1996) Capillary electrophoresis for clinical problem solving: analysis of urinary diagnostic metabolites and serum proteins. *J. Chromatogr. Biomed. Appl.* **683,** 55–65.
4. Neususs, C., Pelzing, M., and Macht, M. (2002) A robust approach for the analysis of peptides in the low femtomole range by capillary electrophoresis-tandem mass spectrometry. *Electrophoresis* **23,** 3149–3159.
5. Moini, M. (2004) Capillary electrophoresis-electrospray ionization mass spectrometry of amino acids, peptides, and proteins, in *Capillary Electrophoresis of Proteins and Peptides* (Strege, M. A. and Lagu A. L., eds.). Humana, Totowa, NJ: pp. 253–290.
6. Kaiser, T., Wittke, S., Krebs, R., Mischak, H., and Weissinger, E. M. (2004) Capillary Electrophoresis coupled to mass spectrometer for automated and robust polypetide recognition in body fluids for clinical use. *Electrophoresis* **25,** 2044–2055
7. Kaiser T., Kamal H., Rank A., et al. (2004) Proteomics applied to clinical follow up of patients after allogeneic hematopoietic stem cell transplantation. *Blood* **104,** 340–349.
8. Neuhoff, N., Kaiser, T., Wittke, S., et al. (2004) Mass spectrometry for the detection of differentially expressed proteins: a comparison of surface-enhanced laser desorption/ionization and capillary electrophoresis/mass spectrometry. *Rapid Commun. Mass Spectrom.* **18,** 149–156.
9. Wittke, S., Mischak, H., Kolch, W., Raedler, T. J., and Wiedemann, K. (2005) Discovery of biomarkers in human urine and cerebrospinal fluid by capillary electrophoresis coupled mass spectrometry: towards new diagnostic and therapeutic approaches. *Electrophoresis* **26,** 1476–1487.
10. Wittke, S., Fliser, D., Haubitz, M., et al. (2003) Determination of peptides and proteins in human urine with CE-MS—suitable tool for the establishment of new diagnostic markers. *J. Chromatogr. A.* **1013,** 173–181.

11. Theodorescu, D., Fliser, D., Wittke, S., et al. (2005) Pilot study of capillary electrophoresis coupled to mass spectrometry as a tool to define potential prostate cancer biomarkers in urine. *Electrophoresis* **26,** 2797–2808.
12. Johannesson, N., Wetterhall, M., Markides, K. E., and Bergquist, J. (2004) Monomer surface modifications for rapid peptide analysis by capillary electrophoresis and capillary electrochromatography coupled to electrospray ionization-mass spectrometry. *Electrophoresis* **25,** 809–816.
13. Liu, C. Y. (2001) Stationary phases for capillary electrophoresis and capillary electrochromatography. *Electrophoresis* **22,** 612–628.
14. Belder, D., Deege, A., Husmann, H., Kohler, F., and Ludwig, M. (2001) Cross-linked poly(vinylalcohol) as permanent hydrophilic column coating for capillary electrophoresis. *Electrophoresis* **22,** 3813–3818.
15. Kolch, W., Neususs, C., Pelzing, M., and Mischak, H. (2005) Capillary electrophoresis—mass spectrometry as a powerful tool in clinical diagnosis and biomarker discovery. *Mass Spectrometry Reviews* **24,** 959–977.
16. Ullsten, S., Zuberovic, A., Wetterhall, M., et al. (2004) A polyamine coating for enhanced capillary electrophoresis-electrospray ionization-mass spectrometry of proteins and peptides. *Electrophoresis* **25,** 2090–2099.

Index

H

I